Five Generations

Stories From My Father

Ronald J. Aubrey,
Retired US Navy 2003

1

Five Generations

Stories From My Father

Ronald J. Aubrey,
Retired US Navy 2003

This publication is intended to provide accurate and authoritative information in regard to the subject matter covered. The statements and opinions expressed in this book are those of the author. The author has tried to be as accurate in historical details as possible, but since these stories are related *as he remembers* them as told by his father, they are subject to inaccuracies.

ISBN-13: 978-0-9964089-1-2
ISBN-10: 0996408916

Proudly printed in the United States of America.

Home Crafted Artistry & Printing
1252 Beechwood Avenue
New Albany, Indiana
www.HomeCraftedArtistry@yahoo.com

ABOUT THE AUTHOR

Ronald J. Aubrey served in the United States (U.S.) Navy for twenty years from 1983 until his retirement in 2003.

During his active duty, he was stationed onboard the USS America, followed by a second sea service tour with Special Warfare Group Two at Little Creek Amphibious base in Norfolk, Virginia. He was assigned as Navigator, then Assistant Boat Captain on the 65 foot Sea Spectre Class Patrol Boat.

After a shore tour at Dam Neck Virginia, he was stationed at Naval Station Sasebo, onboard the USS Belleau Wood. Then he was stationed in San Diego, California. He finished his career at Norfolk Naval Station, assigned to augment the base security as day shift commander.

During his tour onboard the USS America, he participated in the Raid on Lybia on April 15, 1986.

While at Special Boat Unit twenty-four, he was deployed to the Arabian Gulf for anti-mining and patrol duties. Paul Evancoe—Navy Seal was his commander.

Ron is retired and lives with his wife, Devin in Louisville, Kentucky.

During his naval career, Ron qualified as a Combat Craft Crewman.

His awards include:

The Navy Achievement Medal,

The Good Conduct Medal,

The Presidential Unit Citation,

The Navy Unit Citation and

The Meritorious Unit Citation

in addition to various lesser awards.

ACKNOWLEDGEMENTS

I wish to thank those family members and colleagues that supported me in development of this work over the last two years.

Most importantly, my nephew, Herman Joseph Bohn, Jr. (Joe), who is finishing his doctoral degree and provided critical early stage narrative development and book design support for the start of this work (and he would have made one Hell of a Navy officer if we had gotten him enlisted 25 years ago!)

In addition, I greatly appreciate the support of my retired Commanding Officer, Paul Evancoe for getting my young green ass into and out of the Arabian Gulf in one piece. I must also mention the instructors in all of my Navy schools; yes, I used everything they taught me.

My lovely wife Devin, who has never stopped believing in me and has made my life a wonderful thing.

My sister Debbie, for showing me it can be done.

And my son Joshua, who is now a United States Marine; and all of the Aubrey men who have kept the stories alive over the generations.

Table of Contents

Five Generations

Stories From

My Father

Chapter 1. Introduction

BOOK OVERVIEW

I am not an overly religious man. I do believe that God stands with soldiers, sailors, airmen and Marines. He was with me as I travelled into harm's way; that's the only way I can explain my coming home. It is true; there are no atheists in the foxholes.

Even though most people think of the Navy as large grey ships, there is another part of the Navy I was blessed to serve in. Special Operations (Spec Ops), is a world closed to outsiders. The answer to most questions is, "That's classified."

Where you go, whom you see, what you do is in support of national security and defense mission. If you want to do it, but don't want the world to know it was the United States (U.S.), you send in Spec Ops. No headlines, and the glory is kept in house. See you on the gunline.

Now, ask any family, they all have family stories. Some are funny and some are embarrassing for one or more family members. Some can never be told except in a whisper.

These are the stories of the history of our family that my father told me and to which various other family members contributed. This is only the male line, nothing wrong with the female line, but I don't have those stories.

These stories were told father-to-son for five generations, and counting. They are stories of strength and personal conviction. They are stories of card cheating and loosing army jeeps in poker games. They are stories of good men who chose to stand and protect others when they could have hidden and saught safety. The one connecting factor is that they all teach a lesson about how to be an honorable man. The urge to help others and to do our duty is strong in my family. Some people run into burning houses to save others. Some watch in horror as bad things happen. This is a compilation of stories from a family that has stood up when called upon, whatever the situation required.

Enjoy these stories and the lessons in them. If you don't know your family stories, perhaps this will serve as an inspiration to capturing and building upon them in time.

ENSUING CHAPTERS

The ensuring chapters first provide a chronology of the family history as five generations serving in the U.S. military. Each chapter, (2-7), provides brief accounts of recanted events during the service time of each person. Following Chapter 7 is a collection of short stories developed and shared with my son, Joshua Aubrey, during his time in boot camp for the U.S. Marine Corp, of which he is in his 6th year of duty in 2013.

The concluding chapter provides two final stories, developed fictionally, but in part, based upon experience gained during my own time with the U.S. Navy.

Enjoy the material, and your feedback is welcomed.

--Ron

Chapter 2. Robert Whaley

1840-1930

Robert Whaley was an American serviceman
who lived from 1840-1930.

After his service for the Union during the Civil War he headed west to seek his fortune and adventure. In his later years he finally settled in Taylorsville, Kentucky where he met and fell in love with Grandma Whaley. They lived happily until Whaley passed in 1930. He entertained his young grandson Earl, with stories of his travels and adventures. The stories have been passed down through the generations and are retold here.

REVENGE OR JUSTICE

Whaley laid in the desert; he was dying. He had come this far seeking the man who had wiped out the settlers in the wagon train he was leading. A wagon train leader is responsible for the safety of those whom he is guiding. His honor rests on their lives. Whaley meant to reclaim his honor by killing the man who had committed this horrible crime.

He had been out ahead of the wagon train scouting for water when a rifle shot rang out. He had been struck a glancing blow, just enough to knock him out. It was meant to kill him.

When Whaley got back to the wagon train, he was horrified. Men, women, children, all dead. Their supplies were taken; it was a tragedy. Some of the bodies had arrows in them; those persons had also been shot. Whaley looked closely. He could identify most arrows as being made by one particular tribe or the other. These arrows were shoddily made and no self-respecting Indian would have used them.

White men had made these arrows. It was a poor attempt to shift the blame to people who had not committed this crime. Whaley put three of the arrows in his saddle bag and headed for the nearest Army fort.

As he rode, he thought of hatred and vengeance. He thought of killing.

The commander of the fort had seen this work before. He had a name for the man who had committed these crimes, the name was Sidewinder. He agreed to telegraph all forts in the area to be on the lookout. The Army was looking to hang him, or them. Whaley allowed as to how the army could hang the man, if they found him before he did.

Whaley began tracking Sidewinder; murder was on his mind. The trail went into the great desert. Two days into the desert a rifle shot went through his canteen. He continued on. Two days later he let his horse run and started walking. The heat in the desert was murderous. Whaley walked for another day and then he fell.

John Williams, Texas Ranger found a riderless horse just inside the green land west of the great desert. A riderless horse with full saddle and rifle in boot meant only one thing, someone in the desert was in great danger. The Ranger tied the horse's bridal to his own horse's saddle and followed the horses trail into the desert. He found Whaley unconscious, barely alive. He wet the thin man's lips; and when he got a response, he set the man up and got him to drink a little water. Delirious the man mumbled something about a snake. The Ranger checked the man for snake bites but found none. He then threw the poor wretch onto his horse, tied him in place and headed out of the desert. To have any chance to survive, this man had to get out of the heat.

"Easy on that stew," John Williams said to Whaley. "You eat too fast, you'll just bring it back up."

Whaley explained to the Ranger that he needed to kill a man and why. That was when Whaley first learned the philosophy of the Rangers. Justice, not personal animosity was their guiding principle. Rangers stood for the law, protected the weak, and brought wanted men to justice. Rangers worked mostly alone, sometimes, but not very often with other Rangers. The pay was poor, but they got to keep the bounties. If they died in the line of duty, other Rangers would be at the funeral and they would be remembered forever.

Whaley thought for a while and said quietly, "The law, sounds easier on the soul than revenge."

John interjected an important point. "You go where headquarters sends you. If you get lucky, they'll send you after this man you want. If not, eventually they'll send someone else."

"What do I have to do to join up?" Whaley asked.

Whaley was sworn in on the spot. He was given a Rangers star and the turmoil in his soul settled down, slightly.

Whaley would be sent many places over the years; he would bring many outlaws to justice; some would even make it to court. But the prospect of hanging led most to go for their guns and trust their luck. None of them turned out to be that lucky. The next part of Whaley's life had begun.

JIM AND WHALEY MEET FOR THE FIRST TIME

Papa Whaley had told young Earl to go fetch him a stick, and not just any stick, but a very particular stick. It was to be straight, as long as the boy was tall, and three of his fingers thick. It was to be freshly fallen with no rot or obvious cracks. Always keen to help his grandfather with his often fascinating projects, he ran all the way to a nice wooded area. Just to be careful, the boy selected three good-looking branches and brought them back to the old man.

Whaley, often would take an opportunity to teach his grandson odd bits of wisdom and history; so it was today. He carefully examined each of the branches and explained to the boy the characteristics, desirability and flaws of each. Having selected his favorite, he set to work making a cane. He was not as young as he used to be, and at times, his knees would pain him. As he worked, he began to tell a tale.

Travelling by yourself can be a lonely business. But when the Lord figures it's time for you to have a companion, he will provide one for you. It was the end of a long day on the trail and Whaley looked forward to a hot bath, a close shave and a real bed. He had some money, and so far, those few men who had tried to steal it from him, had learned that it was at best, a very bad idea.

The Sun was low on the horizon as he rode into a small town. He saw three cowboys ahead of him just tying their horses to a rail outside of the saloon. Whaley always preferred to give every man the benefit of the doubt and assume they were good men; after all, they tended to prove the quality of their character in short order. So it was with those three.

A young black man in his early twenties, was walking down the boarded sidewalk and the three men began to harass him. Whaley calmly dismounted and hitched his horse to the next horse rail over as the cowboys dragged the young black man into an alley. Whaley slipped the leather restraining loop off the trigger of his six shooter and walked over to where the commotion was going on. He never could stand an unfair fight, and three against one was pretty unfair.

As Whaley came into view of the alley, he saw two of the cowboys were holding the black man and the third was beating him. Then, like the voice of God thundering out of Heaven, Whaley hollered, "That'll be enough!"

The man doing the beating turned and yelled, "Mind your own business!"

Whaley calmly said, "Mister, I just made this my business."

The cowboy laughed as he drew a very long knife, and said, "Then I'll just have to cut you."

Fast as lightning, Whaley drew and fired. The knife went flying and the cowboy dropped to his knees from the pain, holding his right hand in his left. His two friends let go of the young black man and started to back up. Whaley pointed his pistol at them and said, "You two get him to the doc. He's gonna need that hand looked at." Then he looked at the young black man, as the three cowboys hurried away, and said, "Guess we oughta ride out of town. Those three will be back to cause more trouble if we stay."

"Name's Jim, Mister. Thank you for help, but I got no horse."

"Name's Whaley, I got one you can ride." Whaley said as they walked out of the alley. He pointed out one of the cowboys horses for Jim to ride. They saddled up and quickly rode out of town.

They passed the town limits and Jim said, "Thanks for letting me use your horse, Whaley."

"Oh, that's not my horse." The tall man said, "It used to belong to the man who was beating on you. I figure he lost rights to it by his poor behavior."

As they rode on, Jim laughed uproariously and took inventory of his new belongings. He had one horse, one saddle, two saddle bags with food and camp equipment, a blanket, and best of all, a double barreled shotgun with ammo. The last was the best. Jim, as it turned out, was a poor shot with a pistol, as he had never owned one.

Whaley lost a chance for a close shave, a hot bath and a real bed, but he had gained a friend. All in all, he figured he came out ahead.

Not a bad day at all.

WHALEY MEETS A GHOST

It was bedtime. The young boy in the bed looked up expectedly at his father and said, "Daddy, can you tell me another Papa Whaley story?" His father smiled and said, "Whaley was a man who wandered the west after the Civil War. He was a tall, spare man. Spare is an old fashioned word for thin, but he was all muscle. He was unfailing polite. He said Sir and Ma'am. He would tip his Stetson hat at ladies he passed on the street. When formally being introduced to a lady, he would remove his hat entirely and make a charming comment. Whaley was a kind man, slow to anger and lightning fast on the draw—when the situation required gunplay. He was at times a wagon train leader, a Texas Ranger, a survivor in any environment, and he could shoot the fly off of a buffalos butt at a hundred yards and not even bother the buffalo."

An old man and a young boy walked across a field in the country; it was sometime in the 1930's. They walked towards a house in the distance. As they walked, the young boy, whose name was Earl said, "Tell me another story about the old days, Papa Whaley."

"Goodness, boy," The man named Whaley said, "will you never tire of an old man's stories?"

"Never Papa!" The young Earl cried.

"Well, all right." Whaley said. "Did I ever tell you about the night I spent in camp with a ghost?"

Forty years earlier Whaley, and a slightly-younger black man named Jim, travelled the West together on horseback. Towards the end of one particular day, they came to a small hill, nicely wooded, with water nearby. Whaley and Jim fed and watered the horses, then set up a simple camp. A simple camp in those days was a blanket made into a lean-to. They made a fire and cooked up beans with the last of their bacon. Dinner finished, and metal plates and pot cleaned at the creek, there was still a little light left.

Jim looked back the way they had come and said, "I saw some rabbits at the bottom of this hill. Seeings as how I don't much care for just beans for dinner I'm gonna set a couple of traps down there, and see if I can get us some meat for tomorrow."

"Good enough. " Whaley said, "I'll get some more wood for the fire."

Their self-appointed tasks completed, the two friends sat and chatted around the fire. After a time, from the base of the hill in the direction they had been travelling, they heard a man call, "Hello, the camp."

In those days it was considered very impolite and possibly quite dangerous to walk up to a stranger's camp without announcing yourself.

"Come on in, friend." Whaley called back.

Two middle-aged men walked up and the taller of the two introduced himself, "I'm John Jensen, and this here's Bob. We saw your fire and thought you might want to come on down and spend the night at the farm."

"Well that's awful nice mister, but we've already set up camp and I believe I'll stay right here. But we'll be pleased to come down for breakfast." Whaley told him.

Jim agreed. Then the man who had done the talking said, "Well, you can do as you wish, but I should tell you that this hill is haunted by a very violent ghost."

Jim started packing his things at once. Whaley looked at him and said, "Where you going, Jim?"

"Anywhere but here. I don't stay where the dead are. Come on with me."

Whaley simply leaned back against a tree, moved his hat to cover his eyes and said, "I like this spot. I'll be down in the morning." No amount of argument could get the tall man to leave his camp, so Jim went with the other two men to spend the night at their farm, which they assured him was not haunted.

That night, on the hilltop Whaley heard a loud voice, "Leave this place now!"

Whaley, not being one to run from pretty much anything called back, "I think I'll stay right here if it's all the same to you."

Lightning flashed followed by a great clap of thunder, again the voice louder, "Leave now!"

Whaley got to his feet, "Show yourself and we'll see who leaves first."

A tall, heavily muscled Indian came out of the shadows and pointed in the direction the farmers had gone with Jim.

Lightning and thunder flashed again. The wind started to blow wildly. Whaley did not move. Then, with amazing speed, the Indian closed the distance between them. Before Whaley could react, he was thrown twenty feet. Whaley got back to his feet and put a hand to his bloodied lip and said, "Alright, we'll do it your way."

Down at the farm, Jim got little sleep as lightning flashed, and thunder rang out most of the night. Just after day light, as the two farmers and Jim were washing up for breakfast, Whaley came riding up on his horse. His face was bruised, his clothes were torn, and he was a little slumped over. The men watched him as he got off his horse and splashed some water out of a bucket on his face.

"Lord, Mister, "John said, "What happened to you?"

Whaley looked at the man and said, "I met your ghost. Turns out that hilltop is an Indian grave yard. I put up a couple of crosses so folks will know and treat the place with the proper respect."

"Okay," John said, "But what happened to you?"

"Well, we had a little difference of opinion before he told me it was a grave yard." That was all that Whaley ever said of the matter.

THE RAILROAD MAN

Jim was proud of Whaley. The tall man had actually made money playing cards and hadn't even cheated.... well, sort of.

They had just come into a dusty, little cow-town called Tucson. While Jim got the horses stabled, Whaley had gone to the telegraph office as he usually did whenever he was near one. Being a Texas Ranger, he was obligated to keep headquarters appraised of his where-a-bouts. The Rangers often got a lead on where one or another wanted man might be and whichever Ranger was in the vicinity, they would send him there to check it out. The Rangers policy was, "He can come in nice and peaceful like, or he can go for his gun. If he goes for his gun, pay to have him buried." The Rangers didn't see a need to plan for failure.

Jim finished with the horses and stopped by the boarding house to arrange for a room for the two of them.

When you came in off the trail, it was a nice indulgence to get a close shave and a hot bath. Best of all was a real bed. The trail was dusty, and the ground was hard. Having finished his self-appointed duties, Jim walked over to the town's Saloon. Jim rarely drank and Whaley only drank sparingly.

Jim walked up to the bar and ordered a beer. Jim did not care for gambling. He didn't mind if others played but he didn't feel it was proper for him to give away his hard earned cash, or to take it from others. He saw that Whaley already had a man at the table and the cards were being dealt. Jim knew that Whaley had paid for the bottle and would share freely with anyone who sat down to play cards. Whaley had summed up his feelings on the subject to Jim one time, "A man who chooses to play cards, is playing cards. A man who drinks too much, drinks too much. But a man who drinks too much while he's playing cards, that man is just giving his money away, and I might as well get my share."

Jim nursed his beer slowly as two more men joined the game. Cowboys—by the looks of them—sitting at the table, more for the free whiskey than for the card playing. Jim had seen this before. Often, when Whaley lifted his glass, he merely touched it to his lips. It wasn't dishonest but Jim could see it was as close to dishonest as one could get without actually lying. Whaley called it "setting the mood."

Winning hands went to each player pretty much randomly. Whaley folded more than random chance might predict. He wanted to keep the game and the liquor going. He would make his money later.

A large portly man, (portly being an old fashioned word for heavy), joined the game. He was well dressed with a pocket watch and a bowler hat. The usual pleasantries were exchanged. He was a railroad man in town on business. He introduced himself as Mr. Carver. Whaley remarked that he hadn't seen any railroad tracks nearby. Whaley had a disliking for—and great mistrust of—the railroads. They tended to use unfair means to acquire land for their tracks. Mr. Carver allowed as to how the railroad had many interests all over the country.

Jim chatted idly with the bartender to pass the time. It turned out that there was to be a bank auction the next day. The woman who owned and ran a nearby orphanage had fallen behind on her payments and the bank had no option but to foreclose on the property. The rumor was that the cash she had for the mortgage payment had mysteriously disappeared shortly after the Mr. Carver had gotten into town.

Whaley slowly drained the pockets of all the men at the table. Eventually, it was only him and the railroad man. The stakes ran very high, and every time the railroad man lost a hand, he got more and more vocal. Whaley took pity on the man after a while and said, "The cards don't seem to like you today mister. How about I buy you a drink at the bar?"

Whaley had told Jim once, "When you gamble, never beat a man so badly that he leaves with a grudge against you. That only causes problems."

Whaley and Jim had dinner that evening in the hotels dining room. They discussed the news of the day. Whaley expressed an interest in going to the auction the next day. He looked across the dining room at the inebriated railroad man and said, "That could get interesting." Jim had been at ease all day, but something about the tone of Whaley's voice hinted at trouble. Jim was used to trouble; it seemed to follow where ever Whaley went. That night, Jim cleaned his double barreled shotgun and checked that it functioned properly ...just in case.

The two friends were up early the next morning. After breakfast, and a close shave at the barbers, they walked to the sheriffs' office where the auction was to be held.

There was quite the crowd for such a small town. Standing a few steps from the Sheriff was a pretty young lady who looked quite sad. "That would be the lady who owns the orphanage." Whaley observed.

"Too bad she is about to lose it." Jim commented. "Just don't seem right to sell out an orphanage. What's gonna happen to the children?"

Whaley commented slowly, "I imagine they'll sleep at the orphanage tonight." Jim gave his friend a long questioning look. All that he got in the way of a reply was that mischievous grin that told him Whaley had a plan. Jim wondered if it would be a good idea to go get his shotgun, but decided against it. A shotgun is a bad weapon to use in a crowd. Jim just reached down with his right hand and made sure his Bowie knife was secure in its sheath.

They watched the railroad man, slightly hung over from the night before, chatting with the bank president. Behind the railroad man, were two scruffy-looking, slightly overweight men, just watching the crowd. "Hired guns," Jim thought to himself. Jim casually walked around the edge of the crowd. He was completely unnoticed as he placed himself behind the two gunmen. He waited.

The Sheriff began the auction with the required statements about overdue mortgage, no possibility of payment, and full cash payment to the bank at the time of auction. The bidding got under way. A few of the town's citizens had put together a fund to try to save the orphanage. They had put together enough for a fair price but were soon outbid by Mr. Carver.

The Sheriff who hated this part of the job, paused a bit longer than he really should have, and said, "Bid is seventy-five. Going once. Going twice."

From the middle of the crowd Whaley called, "One hundred dollars, cash." Whaley, a careful observer, knew that the railroad man did not have a hundred dollars. The tall man had relieved him of it last night at the card table.

Mr. Carver, seeing who was bidding against him called out "One hundred fifty dollars. Letter of credit."

The Sheriff reminded the crowd, "This is a cash sale, sir."

Mr. Carver, a desperate man, knew if he didn't buy this land he would lose his job, called out, "This man is a thief! He stole that money from me last night."

Whaley moved to the front of the crowd, ignoring the Sheriff he looked dead into the eyes of the railroad man and said—loudly enough for the crowd to hear, "As I recall, I won that money from you playing five card stud." The crowd giggled.

The barkeep who was in the crowd hollered, "He's telling the truth. They played in my bar. And that Carver feller can't hold his liquor." The crowd burst into laughter.

The now very desperate, and outraged Mr. Carver looked over his shoulder and said to the gunmen, "Boys, take him."

As Mr. Carver looked over his shoulder, Jim stepped right behind the two gun men and easily lifted their pistols from their holsters. "Don't move." He said quietly. They were smarter than they looked. They didn't so much as twitch.

The Sheriff continued, "Going once, going twice, sold to the tall man in the white shirt."

Whaley stepped forward and introduced himself to the Sheriff, handed over the money and turned to the woman who used to own the orphanage. She was on the verge of tears. As Whaley signed the Deed, he said, "Madam would you do me the honor of signing this piece of paper?"

She looked at him strangely and said, "Sir, my signature is not required."

Whaley looked at the Sheriff and asked, "Doesn't she have to sign the deed for the property to be hers?"

The Sheriff, a wise man who could think quickly on his feet, replied, "Yes, but also money has to change hands."

"Very well," Whaley said. He turned to the young woman and said, "Would you happen to have a silver dollar Madam?"

She reached into her purse, in a state of shock and handed Whaley a silver dollar.

"Thank you so much. Could you sign here, please?"

She quickly signed the document, then looked at Whaley and asked,"Why have you done this, Sir?"

With a quizzical look on his face, Whaley said, "Just seemed like the right thing to do."

As the young lady jumped up and hugged Whaley while crying and thanking him, God, and Jesus, Jim emptied the two revolvers and put them back in the men's holsters. He said quietly, "Have a nice day." Before they could turn around, Jim had melted into the crowd.

As the two friends walked back towards the hotel, Jim asked Whaley, "We got anything left, besides a real good feeling?"

"Yes we do." Whaley replied, "We've got twenty dollars and some change, and oh, yes, that good feeling."

Whaley said he needed to stop by the telegraph office. Jim said he would go with him. As they passed their horses Jim slipped his shotgun out of its place behind the saddle and carried it barrel down. ..just in case.

RICH AT LAST

Whaley and Jim decided it was time to make some money. The bounty on the occasional wanted man was enough to keep them going, but they had agreed that it was time to get up a little nest egg. So, they trapped their way down the Mississippi. They had a fine supply of animal pelts, and halfway to New Orleans, they had to buy another horse to carry the entire load. If not rich, they knew they would be at least well off.

They met other folk along the way, and if one or two of them thought of thievery, that idea was quickly dismissed from their mind when Whaley deliberately put on a trick shooting demonstration. The two men arrived in New Orleans and sold their pelts for a good amount without incident. With the money, Whaley and Jim decided to buy some fancy gentlemen's clothes.

Whaley looked quite dapper in his fancy three-piece suit and new Stetson. Jim went with a light brown suit and a bowler. They dined at the finest restaurants. They stayed at an expensive hotel and booked passage on a stern-wheeler headed north.

Gentlemen of leisure at last, they even attended services at a local church. Jim met a very lovely Mullato lady of means by the name of Dominique, who was also headed north, and they spent quite a bit of time together. New Orleans was, and still is, a very cosmopolitan city. It is part Spanish, part French, and with all manner of persons from various parts of the world, it constituted a major port city.

When the day came to depart from New Orleans, Jim escorted his newly-met lady-friend up the gang plank. The three of them ate together in the steamboats dining room and commented on the wonderful cuisine. Dominique compared it very favorably to some meals she had had in Paris, France. Whaley and Jim had done their own share of travelling, but had never been out of the country. The three retired to their separate rooms at 11:00 that evening.

Jim was strolling the deck the next morning arm and arm with Dominique, when he saw Whaley playing cards with several large working class men. "Oh, no." He said, "Oh no, no, no."

Dominique commented that Jim had turned almost white and inquired what was wrong. Jim knew that in these circumstances, Whaley would most likely find a way to cheat. The bad part was, that he didn't cheat that well, and got caught more often than not. The two men had had to leave more than one town in the middle of the night over such behavior. Jim also knew that on this boat there was no place to go.

Jim made his excuses to Dominique and tried to work his way over to Whaley and get him out of there before something bad happened. He was six feet away when it happened. It went very fast. There was yelling and fists flying. Money was pulled out of Whaley's pockets and then, he was thrown over the side into the river. Jim ran to the rail and laughed at his friend. Then Jim heard from behind him, "Aint you with him?"

Jim wagged his finger at the three men as he leaned against the rail and said, "There aint enough of you to throw me off this river boat." Then he dived over the side, joining his friend in the river.

"Dang it Whaley!" He yelled when he got to his friend, "I told you to cut that out."

"My apologies Jim." Whaley answered, "You were right."

"I was right?!" Jim yelled, "Of course I was right! Don't forget, our money is on that boat."

Jim yelled at Whaley all the way to the river bank.

They settled on a plan to work their way up river to the next town the steamboat was to stop at, hoping to find at least their belongings dropped off.

The sheriff of the river town was quite amused at their story. He gave them all the reasons he should lock them up, but then said since the boat had left already, there was no one to press charges. He had all of their belongings, but the boat crew said they never found any money.

Jim had a couple of hundred dollars on him when he went in the water. Whaley, of course had none, since it had been taken from him before he was thrown over the side. They found a couple of horses and saddles and continued on their way.

ALL AVAILABLE RANGERS

Jim and Whaley had spent a few uneventful days in the small town of San Jacinto. Whaley checked the local telegraph office at the train station daily. He was waiting for his next assignment from Ranger Headquarters. Jim had met a young Mexican lady and they spent some pleasant times together. Whaley passed the time at the saloon. Making a few extra dollars he called it. He was careful to not take too much money from any one man. It was a quiet, pleasant time, which Jim thoroughly enjoyed. It was actually too quiet. Jim had a feeling, something big was brewing.

Whaley walked into the telegraph office and the orders had come. It read: All available Rangers; report to San Martin Stop; El Diablo to attack town as staging area for robbery of passing train gold shipment, Stop. Stop at all costs. Stop. End.

When Jim read the telegram he looked at Whaley questioningly and said, "The Devil is going to rob a train?"

"Looks that way." Whaley replied as he walked to the town's general store. He wanted more guns and ammo for this job. He had heard of the gunman who called himself El Diablo. A medium-sized group of men had apparently gathered around him, in the hopes of loot and riches. They killed indiscriminately; men, women and children. Rumor was, they had even murdered a priest. This was one man Whaley wouldn't mind putting a bullet in. The telegram said nothing about arresting the man. Evidently headquarters didn't think he could be taken alive.

Whaley looked concerned when he told his friend,"Jim, you just might want to sit this one out. It could get a little nasty."

Jim chuckled as he saddled his horse, "Don't really think my people would be too proud of me leaving a friend in a bad spot. Neither would I."

"Well, just who are your people?" Whaley asked. "You never said."

"I was raised Cherokee." Jim answered. "They took me in as a baby after my folks died. I would rather not disappoint them."

Whaley studied Jim hard for a moment and said, "Sounds like there's a young lady in that story, somewhere.

Jim just smiled.

Whaley sighed and reached into his saddlebag, he pulled something out. "If you're gonna get into this…. Raise your right hand and repeat after me."

Right then and there, Whaley swore Jim in as a Texas Ranger. He pinned the Texas Rangers star on Jim's shirt and said, "I'll let Ranger Headquarters know they need to start paying you now."

Jim smiled and said, "You get paid? Heck I thought you did this job just because you like getting shot at."

It was three days to San Martin. The two Rangers travelled at a good clip and went straight to the Sheriff's office. The door was locked. A passing shopkeeper told them that the Sherriff had heard El Diablo was coming and had lit out of town. The only people left were the ones who had no place to go, and no means to get there.

Whaley thanked the man for his information. The whole town was aware of what was about to happen. They knew trouble was coming from the south in just a few days.

As was their custom Whaley headed for the telegraph office while Jim took the horses to the stable. Jim went to a small creek nearby to start making arrowheads. He figured he would need quite a few arrows.

Whaley was at the saloons bar when a heavy set man came up next to him and ordered whiskey. "Morning Whaley" he said.

Whaley smiled and said, "Been a while Christian. How you been?"

"I have been okay." Christian replied, "Matter of fact, I was just fine till I got that telegram."

Christian downed the shot of whiskey the bartender sat in front of him and continued, "This sounds like a bad one."

Whaley laughed and said, "I guess if it was an easy job, they would have sent somebody else."

The two Rangers sat at a table and began playing cards. As the day wore on, two more Rangers joined them. Whaley explained that he had deputized his friend after he refused to be left out of the action. Al, whose name was actually Albert, commented, "If you like him, that's good enough for me."

The fourth Ranger, Durango, looked at the hand he had been dealt and asked, "You still dealing from the bottom of the deck, Whaley?"

Without bothering to look up Whaley said, "I have no idea what you're talking about Durango. But if it'll make you feel any better you can deal." He smiled as he said that.

"That won't do me any good unless you take that Ace out from under your sleeve." As Durango said this he reached across the table and grabbed Whaley's hand, slid his shirt sleeve up and exposed the Ace of Spades Whaley had hidden there.

Whaley smiled as he said, "Now how in tarnation did that get there?" The four Rangers continued playing cards and drinking slowly. They wanted clear heads for the coming action. Whaley didn't really mean to cheat his fellow Rangers, but he did enjoy seeing how long it took them to catch him at it. They were all-familiar with Whaley's tricks, and didn't take it personally.

After a time, the four Rangers left the saloon and headed for the general store to procure additional items for the job. Durango chose two cases of dynamite and all the fuse cord on hand. He had a great pitching arm and loved dynamite. Among the other purchases were two pistols, he already had two pistols, (he figured he would spend less time reloading). Christian picked up two Winchester repeating rifles, (he already had two. More is always better). Al picked up two brand-new Bowie knives, (he was well known as an artist with a blade of any sort).

Now fully equipped, the Rangers headed for the town's diner; rumor had it the food was excellent.

Jim joined the other Rangers at the town's diner for the evening meal. Introductions were made all around, and Christian said, "I reckon somebody ought to go take a look and see where El Diablo and his friends are at."

Jim spoke up, "I'll head out there after supper. My grandfather was a shadow walker. He taught me how to move at night without being seen. No offense, but you guys will just stick out like a sore thumb."

The next morning, Jim met the men for breakfast. "I found 'em camped just across the river, but they have to deal with a mess before they come here. Somehow their supply wagon and their ammo caught fire." Jim just smiled.

Jim continued, "There is a ridge line that runs for several miles just outside of town. On the other side of the ridge, is a dry creek. They'll have to fight their way uphill and that will slow them down considerably."

Christian spoke up, "I saw a train car off to one side of the tracks. Took a little peek inside. The thing is full of barrels of lantern oil. We put that in the creek bed and light it at just the right moment that should improve the odds in our favor."

"I don't think we can afford that much oil, Christian." Whaley commented.

"That's okay." Christian said, "It looks like stolen goods to me. As Rangers we have a duty to confiscate it."

Al gave his input, "I'll go find us a wagon. You gentlemen care to meet me at the train tracks?"

Ranger Headquarters would eventually pay for the oil. But not until after much negotiations, and a good deal of hand ringing and bean counting. It was, after all a large amount of oil.

It turned out the Rangers needed three wagons for all the oil. Ironically, the town folk were somewhat unwilling to contribute their personal belongings towards their own defense. Al confiscated the wagons as "possibly" stolen. The Rangers placed the barrels in the dry creek bed, fifty feet apart, and Durango attached a stick of dynamite to each barrel and ran the fuse cord from one barrel to the next. The Rangers would insure El Diablo and his men tried to cross the creek bed at this spot by offering themselves as bait.

As the sun began to set the five rangers set up camp, and worked out who would stand watch and when. Jim being the newest ranger, got the midnight watch. They expected to see El Diablo and his men the next day.

Christian had the first watch. He took a spyglass from his saddlebag and studied the horizon. He saw dozens of camp fires, and whistled to himself. He thought, "Now that is a whole lot of bandits." They would definatley see action in the morning.

The next morning, the Rangers ate a small breakfast and made their final preparations for the battle. More than one of them said a silent prayer that they would be successful in the coming battle. A line of dust appeared on the horizon. The bandits were on the move.

Whaley lit the large pile of brush the men had gathered the night before. They had deliberately included a good amount of green cactus to make a smoky fire that would be visible for miles. All they could do now was wait.

When the mob of bandits came into rifle range, Whaley said, "Hold your fire." Four of the men were expert shots but they needed every round to count. Jim had hunted large and small game with his favored weapon, the bow and arrow, for his entire life. He was comfortable that today his arrows would find their targets.

The bandits started firing at two hundred yards. Their marksmanship was very poor, and didn't present much of a threat to the Rangers, at that range. At one hundred yards Whaley called, "Open fire!" The bandits dropped left and right as the Rangers used the lever action rifles with great accuracy. At fifty yards Jim began sending arrows into the mass of men and horses. The ditch was twenty-five yards from their position. Christian lit the fuse for the oil barrows but the cord sputtered out just before the first barrel. Jim, thinking ahead, had prepared several fire arrows the night before. A rag wrapped around the arrowhead and soaked in oil would burn very nicely. He lit one of the fire arrows and sent it into the first barrel just below the dynamite.

The barrels began exploding just as the bandits reached the ditch. Christian began lighting and tossing sticks of dynamite, putting down three or four bandits each time. The explosions spread burning oil over most the front part of the mob. A few of the Rangers well-placed shots were meant to provide a merciful end to the flaming bandits. They could not stand to watch men, even men such as these, burn to death. The odds were beginning to turn in the Rangers favor. With the fire and smoke making it difficult to see their targets the Rangers began reloading. Christian was firing from the ground as he had taken a bullet in the leg. Al had taken a round in the left shoulder, and was slowed way down, as it was difficult to fire and reload with just one good arm.

As the oil fire began to burn down, five bandits on horseback jumped across the creek bed, and the rangers switched to their pistols. Whaley's hat went flying after one particularly close shot. "That was my best hat!" He yelled as he shot the offending bandit in the head.

Once across the creek bed, the five bandits turned right, and worked their way up the ridge heading away from the Rangers, who by now, were getting desperately low on ammo. Only fifteen of the bandits were still in the fight as the five horsemen reached the top of the ridge. They turned their mounts and charged the Rangers. Jim pulled his shotgun and shot both barrels into the charging bandits. Three of them fell. Christian, now completely out of ammo pulled out a Bowie knife and throwing it, hit the lead bandit in the chest.

Whaley sighted in on the last bandit and pulled the trigger of his Peace Maker. The hammer fell on an empty round. Christian threw a bowie knife but it bounced off of the man's chest without doing any harm. Whaley pulled his .22 caliber derringer from his vest pocket. It was all the fire power he had left. He fired both rounds, saw them pierce the man's shirt, and yet, it appeared to have no effect. As Jim notched up his last arrow the man laughed and said, "I am El Diablo. You can not kill me for I cannot die!" Then he pointed his pistol right at Whaley and cocked the hammer back.

Jim let his arrow fly, it went through the man's throat slicing through the spine, and he fell to the ground, dead. The remaining bandits seeing their leader fall, turned to the South and left the battle field.

The Rangers looked around themselves; smoke, drifted across the battlefield. They were stunned to still be alive. They began treating the wounded. Whaley, who had taken a bullet through the muscle of his thigh, limped over to El Diablo and ripped his shirt open. Underneath the shirt was very Old Spanish armor.

This was the secret to his claim of being undying.

"Can't die, huh?" Whaley said to the dead man.

SIDEWINDER

It was a nasty hot day, in a nasty little town, just south of the border. The town was so nasty, it didn't even have a church. The mayor had quit, the sheriff had been shot and killed months ago, and nobody else wanted the job. Seeing's as how they were actually in Mexico, Whaley had no jurisdiction as a Ranger there; but he did have reports of the man he was seeking. Wanted in six western states for robbery, horse thievery, kidnapping, murder and various other crimes, but mostly and most personally, Whaley wanted him for shooting up a wagon train of settlers, which Whaley was leading, to what they hoped, would be their new home in Arizona. The Texas Rangers wanted him, and Whaley wanted him, too.

Jim, calm as always agreed to go across the poorly marked border with Whaley. Whaley had saved his life on more than one occasion. They had gotten into and out of various scrapes, often by the skin of their teeth. If Whaley wanted to go, Jim's opinion was that he would go with him. Jim had built up a little savings from the rewards he and Whaley always split evenly. He intended to settle down one day, but that day had not yet arrived. So as always he rode by Whaley's side, ever alert for trouble.

They tied the horses to the rail just down the street from the saloon. Whaley said, "Watch the horses please; I won't be but a minute." Jim had finally taken to carrying a six shooter. And through practice, had become somewhat proficient in its use. But he never went anywhere without his trusty shotgun.

Jim watched Whaley walk into the town's only saloon, and began his own walk in that direction. He had heard about Sidewinder from more than one Ranger. He understood that this man was the most dangerous they had ever hunted.

From just outside the swinging doors, Jim saw the bartender bring up a shotgun. Jim quietly walked in and laid his own shotgun across the bar with his left hand. The barkeep put his hands up and took a step back. In Jim's right hand was his revolver, which he trained to the right, in order to cover the rest of the bar. Jim surveyed the bar.

Whaley approached a table where several men sat playing cards. "Sidewinder!" Whaley called in a firm loud voice. "My name is Whaley, Texas Ranger. You are wanted for horse stealing, cattle thievery, and all manner of vile and despicable crimes; including killing every man, woman and child in my wagon train. Lay down your guns, come along peaceable like and I promise you'll get a fair trial after which we'll hang you with a new rope."

The large fat man that Whaley had addressed this to, stood and stepped away from the table. "Can't say as I favor getting hung, new rope or old."

Whaley's eyes narrowed as he watched Sidewinder for any telltale signs of gunplay. Without moving his hand at all, Whaley quietly said, "I was kinda hoping you would say that. Slap leather you son of a buck!"

Since this was personal for Whaley, it was important that the other man draw first, and so he did. The men watching the scene, would unable to say witch man started moving first. And even though, both pistols were fired, they heard only one gunshot.

For a few seconds, both men just stared at each other, then Sidewinder smiled. He smiled an evil, greasy, gap-toothed smile. Whaley twirled his gun once and placed it in its holster. He then turned and began walking towards the saloon's door. As he walked, he nodded to Jim, and then he heard a large dead weight hit the floor. Jim holstered his pistol and threw some coins on the bar. He said, "You might wanna bury him before he starts smelling too bad." Shotgun training across the bar as he backed out of the room, Jim said, "Ya'll gentlemen have a nice day."

As the two friends walked down the middle of the street, looking neither left nor right, Jim said, "That's it, we just saunter on out of town?"

"Well," Whaley said, "You might check behind us."

Jim glanced to their rear and brought up his shotgun. He fired once to the left, and once back at the bar. In both cases, men with guns in their hands dropped. Jim cracked open his shotgun, extracted the spent shells, and put in fresh ones before closing the weapon; ready once again for instant use.

"Always a good idea to check the roof lines for shooters." Whaley said as he drew his pistol and shot the man leaning out from behind a chimney with a rifle in his hand. He fell slumped over the peak of the roof. Shifting aim to the left, Whaley fired once, shooting through the dry goods store sign, killing the man whose hat had been just showing above it.

A man with a shovel in his hands walked out of the doctor's office. Both guns trained on him and he put his arms out away from his sides to show that he was unarmed. He got to live.

The two men mounted their horses and paused. Jim waited on Whaley, as Whaley looked around and spat on the ground. "We're done here." The tall man kicked his horse into a walk along the death-strewn street. Jim thought it wise to keep shotgun in-hand, and look back once in while the way they had come.

They left the town in silence and stopped at the first town back across the border. Whaley sent his telegram to Ranger Headquarters. It read, "Sidewinder resisted stop. Headed west stop. Whaley." In his reports, Whaley didn't like to use the words shot or dead. Headquarters knew what he meant. As he left the telegraph office, Whaley ripped down the wanted poster for Sidewinder and left it on the ground. Once a man died, he was no longer wanted.

OLD RANGERS NEVER DIE

As it comes for all men, king or commoner, hero or coward, the day came for Whaley when the Lord called him home. If you're lucky, you go out with drama and flare, saving a baby from a fire, or in a gunfight with a bad man who has to be stopped. If you're not so lucky, painful disease takes you young, or an unlucky accident hits you when you're not looking. If you're really, really lucky, you get to pass quietly in your own bed in the night.

Whaley, at the end, was really, really lucky. After a lifetime of adventure, he met a good woman to love and a good family. He had no regrets, and no complaints. He'd had a life that was full; he had done some good in the world. And his conscience was clear.

May we all be so blessed.

Folk came from all around to pay their last respects. The Widow Whaley, as she would be referred to, for the remainder of her life, was busy. She was busy with her mourning, she was busy worrying about the food she hadn't cooked, because the rest of the family insisted she needn't be bothered. But most of all, she was busy with all the people who kept coming. Folks she had never heard of came to offer condolences. So, perhaps she can be forgiven for one thoughtless comment, to a young boy who worshipped the ground that Whaley once had trod. Whaley, generous as always, would surely have forgiven. But the young Earl was crushed. When he was going on about one of Whaley's tales she interrupted him with a dismissive, "Boy, don't you know he just told tales out his head of things that surely never happened?"

The thought that the old man that he loved best, was not all that he said, was such a horrible thought for the boy that he went off to a corner and cried inconsolably. About that time, an old, distinguished-looking, black gentleman arrived. Smooth shaven with graying hair, he carried himself with quiet, assured confidence. Well dressed in a suit, Stetson Hat, and well-shined boots, he walked up to the casket and spoke quietly to Whaley. No one heard his words, but when his goodbyes' were said, he turned and inquired of the nearest person as to whom he should direct his condolences.

Walking up to the Widow Whaley, he introduced himself as Jim White, retired Texas Ranger, and longtime friend of the deceased. The first name Jim struck the widow, and slowly, she came to realize that all of the stories that she had suffered in silence for years, just might be true. Then, she realized the terrible wrong she had done to young Earl. She asked Jim if he would have a few words with the boy, to ease his pain. Jim smiled widely as he agreed.

Jim found Earl crying on a couch alone. "Hello, young man," He said. "My name in Jim. I rode with Whaley for many years. He was a good man and I was proud to call him friend."

The boy's eyes widened, his crying ceased. The past and the stories came back to life in his mind. In short order, Jim was telling the same stories that Whaley had told. Finally, the boy worked the courage up to ask the old, black man, "Were you really raised by Indians?"

Jim laughed and said, "When you refer to Indians, it's polite to go by the tribe's name. Yes, the Cherokee raised me. Good people, strong moral character. Then, the time came for me to go out on my own. I thought I was searching for my fortune, but I found something much more valuable. I found a friend, your grandfather. "

The next day was the funeral. Young Earl saw Jim arrive and the elderly, black man came to stand by the boys' side. Jim removed his hat as the minister said his prayers. After the graveside ceremonies ended, Earl asked Jim if he was coming by the house afterwards. Jim said he had other business to attend to, but promised to see the young man before he left town.

Late that night, Earl was awakened by a shaking. He barely had his eyes opened before Jim said, "Get your clothes on boy. There is something you need to see."

They left the house quietly so as to not awaken anyone, and made their way to the cemetery. Jim stopped in the bushes just a little ways from Whaley's' grave, and put his finger to his lips. He whispered to the boy, "What you are about to see, tell no man until you are my age." Earl nodded his agreement.

Three men on horseback, in dusters and Stetsons, came out of the fog. One of the men led a riderless horse. Earl could see silver stars on their chests. He understood, these men must be Texas Rangers. They stopped a short distance from where Earl and Jim were crouched in the bushes. The men faced Whaley's resting place, and the lead rider called out loudly, "Ranger! Ranger! There's cattle rustlers, horse thieves and the like at large. Ranger! Dying's no excuse to just lay there; we still got work to do."

The boy watched in confusion. He did not know what the men were doing.

"Ranger!" The lead rider called. "Mount up!"

From Earl's left, someone stepped out of the fog. He was dressed as the Rangers were. Duster billowing out behind him. Moving quickly, the man got on the fourth horse the Rangers had brought. As the horse reared up, the man turned and looked right at Earl. The man touched the brim of his hat and called, "Rangers ride!" The fourth Ranger bore an uncanny resemblance to a young Whaley.

As Earl watched the Rangers disappear into the night, he heard Jim whisper, "Old Rangers never die. They just go on to help someone else."

The next morning, Earl woke up, not sure if the events of the night before had been a dream, or had actually happened. That evening, as suppertime came and went; his mother was upset that Earls' father was late. Joe arrived finally and apologized. He explained his tardiness by saying that the cemetery owner had needed him, and several other men, to resod a portion of the cemetery grounds where some men had ridden horses the night before. Earl bent his head down and smiled. He also did not give voice to what he was thinking. He knew who the riders were, but could not say, until he was an old man.

Chapter 3. Clarence "Joe" Aubrey, U.S. Army

(1896–1981)

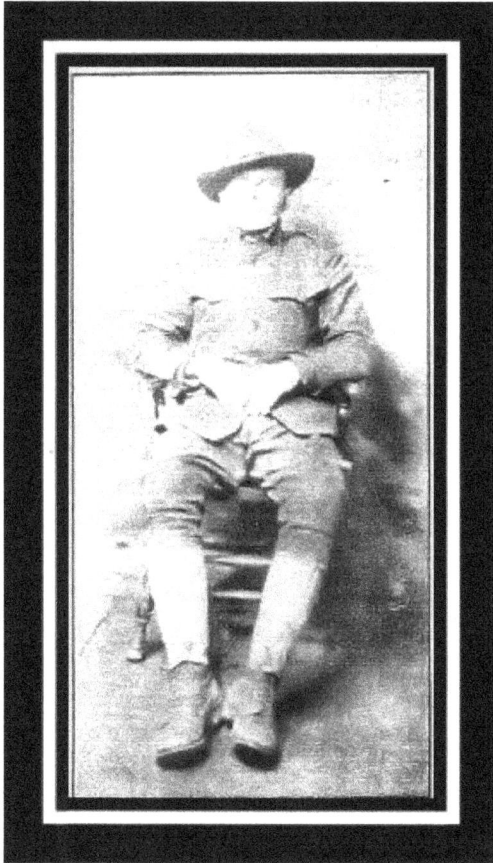

Clarence "Joe" Aubrey, born 1896 and served
Active duty for the U.S. Army in World War I.

Clarence was from a large, Irish family who had a history of land hopping. Back in the 1400's, the *Aubrie's* were originally from Wales. They had titles and land... and an itch to travel. Over a couple of centuries, the family moved to Scotland, Ireland and finally, in the 1800's, to the "land of the free and the home of the brave", America—just in time for the Civil War. When they arrived on America's shores, they changed the spelling to *Aubrey* — to them, it must have sounded more "American."

RIGHT HOOK

I remember my Grandfather, Clarence "Joe" Aubrey, best, as a mountain of a man. So large, in fact, that when he walked, he moved weather systems. He also loved to give little boys a good knuckle rub on the top of their heads.

Joe Aubrey, was a barrel-chested Irishman. He was tall in stature, with blond hair and blue eyes that danced when he spoke. And he was in love with a little German girl—much to her father's dismay.

As strong as an ox, Joe had no problem passing the physical when he went to Enlist in the US Army during World War One. So when Joe enlisted, he had to find a way to get letters to his sweetheart. Conversation between Joe and Mae was discouraged by her family. Yet true love would prevail. They came up with a system to get letters across, "enemy territory". Mae's best friend would also write Joe. And Joe would write her back—along with a few extra pages for Mae.

Life as an infantryman is full of long periods of empty time, during which the men would often provide their own entertainment. Considering Joe's size and strength, his buddies convinced him to be their entry in the unit's boxing match. Joe was also a very smart man. He figured if you get the job done in the first round, you didn't have to go back for twelve more. Now Joe was raised on a farm, and that's not easy work. He was used to wrestling farm animals in order to get them where he wanted them, and this activity gave him a ton of strength.

It was the first blow of the first round. Joe hit his opponent square in the chest with a right hook that had all the power of a landslide coming down a mountain. The man dropped to the ground and never moved again.

Joe was devastated. He was there to kill the enemy, not his friends and fellow soldiers. He never boxed again.

When Joe came home from the war, he went out for one of the first baseball teams in the country, and made it. He had always liked sports, and thought this to be a good fit. Practices went well, and the time of the first game was soon at hand. The crowd was small but boisterous. Excitement filled the air as the men took their positions in the field. Joe threw the first pitch and the batter swung. The ball went right to the pitcher's mound. He caught it with his gloved left hand, and threw the ball with his right hand as hard and as fast as he could to first base.

The ball came screaming towards the first baseman. He got in front of the ball but could not get his glove up in time to catch it. The ball struck him square in the chest. And just like Joe's opponent in the boxing match, the man fell never to move again.

So much for baseball.

Joe went back to Kentucky to do what he grew up doing.

Farming.

Joe thought, "When I kill something there, at least I can eat it."

MOON SHINE BOMBS

The First World War was fought from 1914-1918. After the Americans arrived in their huge numbers in June of 1917, things really got going. The Americans carried the M1903 Springfield Rifle. A bolt action .30-06 caliber fed with a five round clip. Joe Aubrey being raised in the hills of Kentucky, and an avid hunter, was intimately familiar with its operation.

Their squad stopped for the night in a small recently-liberated French town. The town was deserted, the residents having fled the fighting weeks before. During a security sweep, Joe and his buddy Bobby, had found a cleverly hidden still, on a hill just west of town. A dozen glass jugs sealed with corks were already filled, and a fifty gallon drum with a faucet at the bottom was almost full.

Bobby called his friend Billy over, and between them, they squirreled the jars away in their bags. Their Lieutenant came by a few minutes later and declared the moon shine off limits. They had their Medic fill a couple of jars for medical use. The privates did not feel it necessary to let the Lieutenant know about the moon shine they had already confiscated.

After a short and uncomfortable night's sleep, the unit got orders to move out and sweep to the west, to check for German troops. It wasn't long before they found them. The Germans had moved onto the nearby hill during the night and opened up on the Americans just as they hit the tree line. Everybody hit the deck and went for their guns. Joe, Bobby and Billy wound up in the same spot behind where three trees had fallen to create a natural bunker. Joe used his bayonet to cut away some smaller branches to create a snipers peep hole to shoot through. They saw that they were about fifty yards from a German machine gun position. The rest of their squad was off to their left and pinned down, unable to move.

Joe asked his friends for a grenade. They had none. So Joe opened his kit bag and pulled out one of the jars of moon shine.

"Hell of a time for a drink, Joe!" Bobby yelled, as he fired at the machine gun nest with his M1903 Springfield.

"Get yours out too." Joe said, as he pulled the cork off of his jar and stuffed a first aid dressing into the top.

Joe pulled out his last box of matches and lit the now moon shine soaked first aid dressing.

The Germans had a machine gun set up just next to the still the Americans had found the previous day. Joe stood, and with all his strength, threw the flaming jar right into the middle of the machine gun nest. The Germans screamed as they died. Another group of Germans ran up as the flames died down and started working to get the Machine gun ready to go again. Billy worked fiercely to get the cork out of one of his jars. After downing a quick drink, stuffed his dressing into the jug and lit it. He got up to one knee and threw, it fell short.

"Joe!" He yelled, "You'll have to do it. I can't reach that far."

"Load me up!" Joe hollered while he tried to get a clear shot at the nest with his rifle.

"Ready," Billy yelled as he passed his jar to Joe.

Once again Joe stood, valiantly disregarding the personal danger and heaved, with all his considerable strength. The improvised grenade landed just past the nest, and broke against the side of the fifty gallon drum. The faucet was already leaking. Joe dropped back down and grabbed his Springfield.

The Germans had by this time, spotted where the flaming bombs were coming from and zeroed in on the three men. They got a few rounds off and then the fifty gallon drum blew. Germans and machine gun parts blew in every direction! With the machine gun no longer a problem, the American unit was able to beat back the German assault and call in artillery to finish the job.

After the action was over, the lieutenant walked over to the three men and said, "Good work men. By the way, I thought we were out of grenades?"

Joe spoke up. "We were sir. What did you do before the war; if you don't mind me asking?"

"I practiced law in New York." The lieutenant really didn't know where the conversation was headed.

"Well," Joe said. "I guess I can forgive a New York lawyer for not knowing that Moon Shine burns hot and fast." Then Joe offered the lieutenant a pull from an open jug.

"Pass that around, Private. One drink per man, no more."

As the lieutenant walked away, he thought to himself, "No one in New York is ever going to believe this one."

BRASH ACTS OF BRAVERY

The year was 1918. It was a typical late fall day in Fontaines, France—cold and grey. The Great War was raging on for all of Europe. And America has entered into the fray to try and bring an end to the carnage that has enveloped the land for the last four years. On this seventh day of November, courage under fire was the order of the day.

The war had been stalled for months. Thousands of men died in futile charges, which led to advances which could be measured in yards. Trench foot, disease and trauma were common. The advent of industrial warfare had outstripped the tactics of the generals who were still fighting the last generation's war. Until these tactics were changed, the war would continue to be a long drawn out slog of blood and death.

No one had bothered to tell the newly arrived Americans that brash acts of bravery, and personal sacrifice were not only out of style, but also very likely suicidal. It was under these circumstances, that Private Clarence "Joe" Aubrey, along with three other brave soldiers, volunteered to attempt a rescue of several wounded soldiers in no mans' land. They exited their trench and ran across no mans' land, enduring murderous artillery and withering machine gun fire, to try and reach the fallen soldiers. They were driven back by the hail of bullets and shrapnel to the somewhat safe confines of their trench again.

Undeterred, these brave men took up this perilous gauntlet yet another time, facing certain death to reach the stricken soldiers. Forced to the ground to avoid the murderous machinegun fire, Private Aubrey pulled his last grenade from his belt, and heaved it with all his considerable strength at the machine gun nest. The grenade fell short, but did attract the attention of the machine gun crew. The Germans now knew they were being targeted and sought their attacker.

In the smoke and carnage, they missed the big country boy. He yelled at his friends for a fresh grenade. Before the war, Private Aubrey just happened to be the best baseball pitcher in three counties. Using his previous experience, and totally disregarding his personal safety, the big farm boy pulled the pin on his grenade, and came to his full height of six foot three inches. That is one heck of a target. He gave it his best fastball pitch. The Germans manning the machinegun saw the suicidal American, and tracked their fire in his direction. He dropped back to the ground before the bullets found him. The Germans cursed because they missed him, and then saw the grenade land in their nest. They didn't have time to curse again before the grenade exploded. The machine gun nest silenced; the big farm boy from Elk Creek, Kentucky and his friends, advanced and brought the wounded soldiers to safety.

For this action of heroism, Private Clarence Aubrey, along with the three other soldiers, received the Distinguished Service Cross. Clarence also received the French Croix de guerre. The Croix Daguerre was a French medal, which was awarded to foreign soldiers who distinguished themselves by acts of heroism involving conflict with the enemy.

That was my grandfather.

Chapter 4. Earl Aubrey, U.S. Army

(1929 – 1981)

Earl Aubrey was a farmer's son.

And when his Uncle Sam called for a new world war, he answered the call. Earl served in the U.S. Army during the later stage of World War II, and in the post war occupation of Japan (1944-1948). After the war, Earl returned home to Taylorsville, Kentucky, marrying his teenage sweetheart, Dorothy Francis Carlisle, going on to raise four children over the next 33 years.

A FEW WORDS ON GAMBLING

Gambling has a strange history in our family, it doesn't work for us. My father, Earl, was part of the occupation of Japan after the Second World War. He carried his forty-five caliber pistol in a shoulder holster because his unit was out of hip holsters. Never put it past a military police officer (MP) to stop his rounds for a few hands of cards, and a drink or two.

On this occasion, Earl and his partner were on a vehicular patrol, checking other posts and the marketplace. They stopped by a poker game, and out of money, with a great hand, Earl just had to bet their jeep. Predictably, he lost that hand. No one knows what he told the motor pool, but I'll bet it was a good one.

One more gambling story on Earl; family was visiting from out of town, so Earl decided to take them to the horse races. On the first race, coming out of the gate, the horse Earl had bet on fell and broke his leg. In those days they put the animals out of their misery right on the track. Earl said, "That horse died because I bet on him." Earl never bet on a horse at the track again.

I learned my own lesson about gambling at The Paradise Island Casino in the Bahamas. My first ship, the USS America, was making port call. I decided to try my luck at the black jack table. I soon found out that gamblers drink free. It took three days, but I left a months' pay at that casino. Thankfully, I was just a young seaman at the time, so it wasn't that much money.

THE ADVENTURES OF EARL

I knew my father as a hard-working Sears and Roebuck repairman. If it was broken, he could fix it. If it needed to be made, he could make it. There was more to Earl Aubrey than just that. About five foot 10 inches tall, he was comfortable with his country values. He was stockily built and wise.

The stories about my father show the inner strength and fortitude that he was never able to put into words. He literally led by example. When I ran into danger, it was because I saw him do it. When I helped someone who needed it, I had seen him do that, too. He wasn't much on talking, so when he talked, I listened.

A soldier, in the Second World War, he was onboard ship headed for the invasion of Japan when the Atomic bombs were dropped. How many of us wouldn't be alive if that invasion had taken place? So here we are; and now I will tell the tales.

Raised in the country, just outside a little town called Taylorsville, Kentucky, population 875. I am pretty sure the sign outside of town still reads the same. Very little changes in those small towns. It is still much the same as it was then.

Small towns are notorious for not having much for teenagers to do. So there was always a little trouble to be gotten into. Earl got a part-time job as a *ridge runner*. A ridge runner, back in that time, was a person who transported untaxed liquor. This is a homemade alcohol called moon shine. He never got caught, and was a very good driver. You had to be able to get around, and sometimes you had to be able to just get away.

How does a great driver turn a car over on a perfectly straight road coming into town? That is unknown but what is known is, that my father told me, "I felt her come up on two wheels, and as that car flipped, I just laid down across the seat. When it stopped moving I was upside down with the car roof resting on top of the seat. If I hadn't of laid down, I would have been done."

My mother had a little to add to the story. She and her sister Annie, were at the hairdressers when a lady came into the shop and screamed, "There's a man dead in a wreck on the edge of town! I knew he was dead, and he walked out of that wreck dead!"

Thankfully, she was incorrect. When my mother and her sister found it was Earl Aubrey in the wreck, they just laughed and laughed. Earl was known as a good one for trouble. Mom wound up marrying him anyway.

U.S. Army *Range Masters* know one interesting fact. Country boys from Kentucky can shoot like nobody's business. In the country, if you can't shoot, you don't eat. So, at boot camp, when Earl's company got to the range, the Range Master called for anyone from Kentucky to step up to the gun line. Earl and another young recruit stepped up. "You have ten rounds. Show these city boys how it's done." The Range Master had never seen either man shoot. He was just proving a point.

Ten rounds later, each round frighteningly close to center target, the Range Master went on to say. "You will all be able to shoot like that before you leave here."

The next day on the range there was a problem. The targets were lifted up and down by a man in the ditch behind the targets. The target Earl was shooting at never went down. The Sergeant came up behind Earl and said, "Can you hit the pole the target is on?"

Earl smiled and sent a round downrange. The wood pole split and the target fell. Earl didn't have a problem at the range the rest of the time he was in boot camp.

Jim and Susie were "A Portrait in Black and White". They were entertainers at the local Holiday Inn in Lively Shively. Jim was black and Susie was white. In Kentucky, in the seventies, this was progressive. Unfortunately, not everyone in Kentucky was progressive. Earl and Fran spent many happy Saturday nights in the lounge of the nearby Holiday Inn listening to their music. Jim and Suzie came to our house for meals and just to visit.

It was bound to happen sooner or later. A loudmouth with a few-too-many drinks under his belt, started making derogatory comments one Saturday night. Earl considered Jim and Suzie to be friends, so he went over to speak to the unruly gentleman. No one knows what was said, but if you've ever had to deal with a loud drunk, you know how tricky it can be. Earl assisted the man to his feet, one hand firmly on the man's arm, escorted him out of the building. Earl's only comment when he came back to his seat was, "He decided to leave."

From the bandstand, Jim gave Earl a big smile and a wink. Earl just smiled and nodded in reply. There were no more problems at the Inn.

In no way does the following story reflect poorly on my sister's taste in men.

My sister Debbie broke up with her first boyfriend because he scared her. That night, we were all settled in front of the television when there came a loud crash from the living room, followed by the screech of spinning tires. A brick had been thrown through the plate glass window. My father grabbed his thirty-eight revolver, and him and my brother left quickly. I found out later that my brother Jack, and my father had caught up with the young man. Dad just smiled at him and raised the gun up so it could be seen.

That young man moved to Florida for reasons of his health.

HUNTING THE HUNTER

Even after the war was over, some few will always want to take just a few more shots before they personally lay down their arms. As Military Policemen, it fell to Earl and his partner, David, to sweep up the die hards. They had been sent to Area C, were supply convoys had been taking sniper fire. Earl was from the country, and hunting was second nature. The army had just taught him how to use his skills to hunt men.

The two MPs had been two nights and three days in the hills that overlooked the convoy route. Earl could almost smell the sniper they were hunting. He had picked up trail trash along the way. This guy was sloppy, and tomorrow, it would be up to him if he lived or died. Earl had no intention of killing him. This particular guy was going back to US Military HQ. The intell guys wanted to chat with him. Rumor had it that this guy was in contact with other snipers in the area.

Earl and David finished their coffee and the last of their combat rations, (c-rats). David would go down to the river and catch some fish to give them another day on the trail. Earl would provide over watch while the other MP caught dinner, and served as a tempting target. The sniper would have to shoot into the sun, and more importantly, if Earl saw the glint of sun on glass, he would have the man.

The two men moved cautiously through the underbrush. Their M1 Thompson submachine guns at the ready. Each Tommy gun was loaded with a twenty round stick magazine. Both men had plenty of magazines in their packs. It was late afternoon as Dave approached the river. Earl was just uphill with his binoculars and an M1 carbine, ready to fire on the sniper should he expose himself.

Dave felt himself being watched as he filled the two canteens and set about fishing. He had tied some string to a stick, but the bait he dug up looked juicy, so he had high hopes.

A glint of light, high on the hill, was all Earl needed. He swiveled his M1 slightly up and fired. As the round left the barrel, he sent a little prayer with it. A cry of pain from the hillside, and a large splash from the river occurred almost at the same time.

Earl walked past Dave as he stood up. "Let's go," he said,"I belive I got him."

Dave grumbled about being wet and cold but brought up his Tommy gun anyway.

Earl and Dave slowly made their way up the hill, working first to the left, and then to the right. Earl in front, went down on one knee and brought his closed fist up. Military sign language for stop. The problem was, they could hear an occasional moan, but the terrain made it difficult to tell exactly where the sound was coming from.

The two men changed positions and listened carefully. Finally, it became apparent that they had almost walked on top of the snipers nest, which was covered with a top made of woven branches, interlaced with leaves and vines. If they had stepped on it, they would have gone straight through.

Earl pulled his bayonet from its scabbard, Dave drew his .45 colt semi-automatic pistol from its shoulder holster. It took a minute, but Earl finally convinced Dave to jerk the top off of the four-by-four foot hole, after which he would jump in and take the sniper into custody.

Dave pulled, Earl jumped and all hell broke loose!

The Japanese sniper had been hit in the left shoulder, but he had lots of fight left in him. Dave looked down at a sword fight with bayonets. The Japanese was obviously the better fighter and kicked Earl under the jaw, and he went down hard. Dave pulled the trigger on his .45.

Fight over.

Lesson learned.

Never bring a knife to a gunfight.

Chapter 5. Ronald Aubrey, U.S. Navy, Retired

(1960 – present)

"Damn the torpedoes, Full speed ahead!"

--Admiral David Glasgow Farragut (1801-1870)

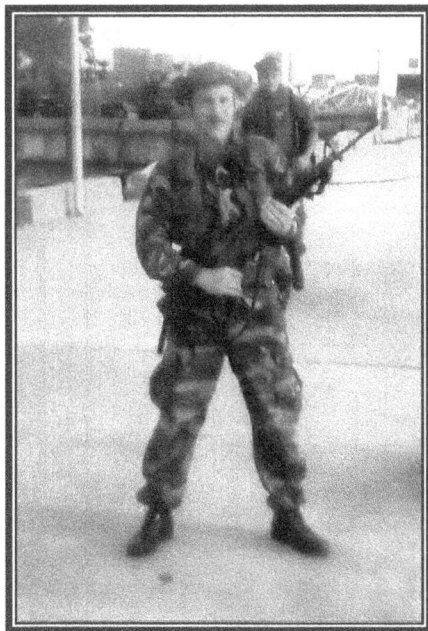

I was born in answer to a promise Mom made God after my sister, Debbie, was born. You see, she was 11 weeks early, very tiny and blue. The nuns came in and told Mom she wouldn't make it.

So Mom prayed.

She told God, if He would save that little baby girl, she would dedicate not only her life back to him, but she would also dedicate the life of her next baby back to the Lord, too, in thanks.

That was where I came in. I was the next baby; and the last. After all, she had only promised to dedicate two children back to God.

It explains a lot.

A BEGINNING

I was a proud little Cub Scout. The time came for the Pinewood Derby. The scouts provided a block of wood which I was to use to make a race car. My father and I worked many hours in the basement workshop to get that car just right. Dad had hand tools, and power tools and tape lines; I mean he had everything. We even found a front bumper, a rear bumper and two pieces of leftover parts from different things to sit in the driver's seat. It was the only trophy I ever won. The best part was making the car with my father.

I went to my father during my first year in high school, I had a bully problem. His advice was simple, "Next time they start, just haul off and punch whoever it is in the nose." I made it my business to follow his direction. The next day at lunch, I came over the table at one of my tormentors; he was the largest football player in the freshman class. I had no problems after that. This was a very strict Catholic school. I could have been thrown out. But Dad went to talk to the principal, and the other young man was not invited back the next school year.

For the next few years, I dedicated myself to enjoying my new found freedom to be unafraid of anyone. I didn't realize you were supposed to be afraid of the cops and the judges. One of the judge's downtown finally got my attention.

Tired of seeing me in court for about the seventeenth time on minor charges, (I was in for possession of marijuana,) he said, "Ron, you know if I put you in jail for a year, I don't have to look at you for a whole three hundred and sixty-five days." I don't recall the rest of the conversation, but the next morning, I started making the rounds of the local military recruiters.

The Army wanted me to be an infantryman. The Marines wanted me to die in glory. The Air Force wanted me to move boxes. Not interested in any of that, I finally went to my last resort. I walked into the Navy's recruiting office at nine o'clock on a Tuesday morning. The recruiter had a neatly trimmed beard, his tie was loosened, his feet were up on his desk, and he was hung over. "Now that," I said to myself, "is a job I can do!"

Seven months later I found myself at the world famous US Navy Recruit Training Depot at Great Lakes.

Upon graduation from Recruit Training, I was sent for my occupational school to Pensacola, Florida. This was Air Crewman Candidate School. Two weeks into training, I injured my knee at the obstacle course. Unable to complete my training, I was sent to the USS America as an undesignated airman. There, I was assigned to catapult number three, the waist cats. I respect the men and women who perform the dangerous job of launching aircraft at sea, but it really wasn't my choice of careers. It took fourteenth months, but I managed to qualify to take the advancement exam for Quarter Master third class. This minor miracle occurred because I spent half my sleep time on the bridge getting qualified as a Helmsman. And most of the rest of my time off studying for the exam.

Now you have to understand that the Navy is divided into segments. Those who work with, or on aircraft, are called air men; they wear green stripes on their dress uniforms. The seamen, those who do primarily shipboard things, wear blue stripes on their dress uniforms. When the Air Department had a personnel inspection, the Department head did not care to have a seaman in his Air Department. I failed yes; he failed me because I was wearing seaman stripes, and had to stand for re-inspection later that evening.

Norfolk, Virginia was where my ship was stationed. I had been the Senior Master Helmsman on the USS America, and I was looking for a different sort of job. Inevitably, a sailor always winds up at a party. It was getting late, and this one young lad was going on about how he worked on a gunboat. The last gunboat that I had heard of sank in the Vietnam River Delta. I felt it necessary to share my small bit of knowledge, and my suspicion that the young sailor worked on a boat transporting guns. Boy, was I wrong. After much discussion, we got in his car so he could show me his gunboat. We entered a guarded compound and drove up to a pier. There she was, (all ships and boats are referred to in the feminine pronoun most likely because it costs so much to keep both of them in paint and powder).

Beautifully spotlighted, with all of her guns mounted, she kind of looked like a porcupine that got all excited. I like to say that the MK3 65-foot Sea Specter class patrol boat carries more guns per acre than a battleship. She had a hydraulically operated 40 mm Bofurs anti-aircraft gun forward, just behind that, a Mark nineteen Mod three 40 mm fully automatic air cooled grenade launcher, a 20mm anti-aircraft gun amidships, (halfway between the front and the back) a fifty caliber machine gun, just behind that, and another 20mm gun aft. You have to throw in two 50 caliber machine guns on the starboard side; one forward of the pilot house and one aft. Add to this, several light machine gun mounts at various places,and you have a porcupine that's all excited. I had to have one! I really didn't know what I was getting into.

I was two weeks married when I arrived at Special Boat Unit Twenty-Four. I was six weeks married and staring out at the stormy waters of the Atlantic from the deck of a Navy amphibious ship. We had been deployed on a few hours notice. We were headed to the Persian Gulf to help insure the smooth flow of oil, during a time when Iran and Iraq were at war.

Halfway across the pond, it was the Fourth of July. Hamburgers, hot dogs and a gun shoot was our celebration. The ship went first. She fired all four of her

50-caliber machine guns, and then it was our turn. The four patrol boats were loaded on the deck, two on the port side and two on the starboard side. They were placed so that the strong port side, the side that had the most guns, was on the outboard side. We all opened up at once. It really did sound like a war. Twenty-four guns of various calibers, shot into the night sky. We had to have been visible for miles.

Certain modifications were made to the topside configuration of some of the boats, also known as PBs for the satellites to look at, but if I tell you about those, I'll just wind up in prison. This was my introduction to the dark and dirty world of Special Operations. Some things you can talk about in public, some things you can talk to your buddies at the unit about, and some things you can't even tell yourself.

Bahrain is a lovely desert city on the edge of the Persian Gulf. A US Naval activity group had a base there at the time. We made port, and got craned off of the ship into the water. We spent a few days doing maintenance that you had to be in the water to conduct. During this time, we were introduced to our new home. It was a new concept to the Navy, it was called a Mobile Sea Base. it was known affectionately to the PB crews as The Battle Barge. Named Hercules, a studly name if I may say so, she was a rented oil derrick barge with a crane large enough to easily lift the boats up and onto the deck for maintenance. It had enough state rooms for all of us and the barges crew with room left over. The housekeepers and cook staff were mostly Pakistani with a few Phillipinos. The food was excellent; I always complimented the "Spinneys" and never skimped on their tips. The name Spinney was somehow derived from the name of the company who staffed the barges. Yes, they cleaned our rooms; as a result we could not have any classified material in there. Since I always took good care of them, they came to like me. And when they cooked Ox Tail soup, they gave me the tail! It tasted wonderful, but I had the worst gas for days.

The plan was to have a tugboat take the barge and anchor out in the middle of the Persian Gulf, and then we would move on board. We got our first operation before the barge even got away from the pier.

DEPARTMENT OF THE NAVY

THIS IS TO CERTIFY THAT
THE SECRETARY OF THE NAVY HAS AWARDED THE

NAVY ACHIEVEMENT MEDAL

TO

QUARTERMASTER SECOND CLASS RONALD J. AUBREY, UNITED STATES NAVY

FOR

PROFESSIONAL ACHIEVEMENT ON 28 AUGUST 1986

GIVEN THIS 9TH DAY OF DEC 19 86

SECRETARY OF THE NAVY

DANCING IN THE STRAITS

It was the fall of 1987. Four months earlier, three United States Aircraft Carrier Battle Groups attacked the country of Libya for its government's poor decision to provide personnel and logistical support for a terror attack on a nightclub in Italy that was popular with Americans. The politicians had all kind of fancy reasons for the retaliatory bombing; for me, it was simple. "You hurt us, we hurt you worse. Oh yeah, don't try that crap again, or we'll be back and make this look like a Sunday school picnic."

After the initial raid, each Battle Group went on to some really wonderful liberty ports in the Mediterranean. After some great liberty, the Battle Groups came from three directions in the middle of the night at high speed. We went charging across the so-called Libyan Line of Death, back to our initial launch points, and then, we turned around and went to another great liberty port.

We did this three times. It's called psychological warfare. It is designed to scare the hell out of the bad guys, and to dig the previous message in even deeper. It worked. After being included in the "Axis of Evil" speech, and watching what happened to Iraq—after the attacks on 911—the Libyan government openly declared its nuclear weapons development program. They also requested international assistance in the disassembly and destruction of that program. That was one less country we had to go to war with.

As the USS America approached the Straits of Gibraltar, it was my job as a Master Helmsman, to take the wheel. A helmsman steers the ship during open ocean transits or during flight operations. A Master Helmsman steers the ship during more complex maneuvering. The time for this training is measured in months; it can only be done on the job while closely supervised by an already qualified helmsman.

The transit is rarely easy and the traffic is never light. All of the shipping going into and out of the Mediterranean (East and West), and from Spain to Morocco (North and South), all come together at this one spot. That spot is only ten miles wide. International treaties define lanes for East/West and North/South traffic but when you put real people in real shipping out there, it can get messy.

The whole Navigation Team for our ship was on station. The Captain, also called the Commanding Officer, (CO) was on the Bridge. He was in overall command and ready to take charge, just in case absolutely everything went straight to hell. The Officer of the Deck, (OOD) in overall charge of the transit stood proudly center stage. He had just finished his qualification as OOD, and it was his first time in charge of a transit by himself. The Conning Officer, or The Conn, stood just to the right of the OOD, and was in charge of relaying engine and rudder orders to the throttle man and the helmsman. The Navigation officer, in charge of the Enlisted Navigation Team, was responsible for relaying the ships position, speed made good, and nearest hazard to navigation, in addition to recommended course to steer to the OOD. The Master Helmsman is responsible for rapidly and accurately responding to orders from the Conning officer or if the Conning Officer was relieved, by the OOD, The Conn will then go to his corner and pout until the OOD tells him to assume the Conn again. The Master Helmsman will then respond to the orders of the OOD.

Finally, if the OOD can't handle the job, the CO will relieve the OOD, who will then call his relief and go to his state room to right his letter of resignation. If the CO can't handle it, the ship is on the rocks and the same Helicopter that brings the new CO, will also take away the old CO, who will then be Court Martialed. In short, if you can't handle the stress of the job, stay home.

I love driving ships. On my duty days I would spend much of my free time reading every ship handling book in the Navigator's library. This greatly enhanced my skills as a Master Helmsman so that on the night that everything had to be perfect, I was ready.

It was the dark of the moon with a stiff wind from the North East as eighty-six thousand tons of aircraft carrier lined up to enter the channel through the Straits. Traffic was extremely heavy, and multiple civilian vessels were ignoring the traffic scheme, causing a hazard to all of the shipping. Once you're in the channel, you're committed. You can't turn around and you can't stop.

Sometimes, you just have to dance with the devil and pray.

The situation on the bridge deteriorated steadily. I watched it happen and understood that the Conning Officer was in beyond his experience. The OOD relieved him, and as additional traffic started crossing in front of us, I saw the OOD loose his situational awareness.

It's not that those two fine officers were incompetent, the amount of traffic and the general disregard of the rules that night would stress and overload any but the finest ship driver. Orders to the helm came fast and furious as we changed course to avoid one contact only to line up on another.

I watched the sail of a sailboat disappear from my line of sight beneath the flight deck at the same time as the CO barked, "This is the Captain! I have the conn! Right thirty degrees rudder."

"Now," I thought to myself. "Now, we're gonna dance."

In order to avoid the shipping in front of us, it would be necessary to turn the ship to the right. Unfortunately, this would send the stern of the ship to the left and sink the sailboat. Ships aren't like cars. The directional force in a car is applied from the front. On a ship, the directional force comes from the rear, or the stern. In order to turn the ship to the right, the stern swings to the left.

The CO had just ordered the ship to turn right. The resultant swing of the stern to the left would sink the sailboat, "Unless," I thought to myself, "the Skipper is going to pivot the ship around the sailboat." A ship when handled properly, will turn around a point at the middle of the ship lengthwise. It's not hard to get right, but when the lives of a smaller vessels crew depend on it, the sweat factor is unbelievable. I knew what he was going to do.

I answered up smartly "Right thirty degrees. Aye Sir!" It takes an incredibly long time for the rudders on an aircraft carrier to transit thirty degrees. At least it does when lives are at stake. I nailed that thirty degrees. It wasn't twenty-nine and a half, it wasn't thirty one, the rudders stopped at exactly thirty degrees. "Sir, my rudders are right thirty degrees." The next step I knew was to shift the rudder to left thirty degrees. I waited what seemed like forever for the next order. The Captain is responsible for his ship. He is also responsible for anyone killed or injured by the ships maneuvering. It was the Captain's responsibility. It was the Captain's call.

Finally, "Left thirty degrees rudder!"

As soon as the CO said left I turned that wheel as fast as I could, while answering up "Left thirty degrees. Aye Sir!"

The OOD barked to Boatswains Mate of the Watch, "Boats, I want to know when the aft lookout has that sailboat in sight. I want to know its condition!"

Another snappy, "Aye Aye, Sir." Boats sent the information over the sound powered phones.

Another eternity passed and Boats reported, "OOD. Aft lookout reports sailboat in sight to port. No apparent damage."

"Right fifteen degrees rudder." The CO ordered.

"Right fifteen degrees rudder, Aye sir." I said as calmly as I could.

There were no other close calls that night. And when the ship got into open waters in the Atlantic, I was relieved by the scheduled Helmsman. I took the chance to step into the chart room and light a cigarette. My best friend on the ship, Kevin, a fellow Master Helmsman, came into the chart house, very excited, and said, "Dude! I've never seen anybody drive like that! How did you do it?"

I took a drag on my cigarette and said calmly, "I just followed the Skippers lead." I hoped he didn't notice that my hand was still shaking ever-so-slightly. The Captain may be responsible, but my hands were on the wheel. And yes, I did take a second to silently thank God that no one got hurt. At that point, the CO's orderly, a young Marine, burst into the Chart House and said, "Man! You just got a Navy Achievement Medal! (NAM) The Skipper said he never saw anyone drive like that!"

I looked at Kevin and said, "Is there an echo in here?"

The Marine explained that the Captain had told the OOD to write me up for a NAM. Some sailors take years to get a NAM. I just got one in one night.

Not a bad night's work.

The Secretary of the Navy takes pleasure in presenting the NAVY ACHIEVEMENT MEDAL to

QUARTERMASTER SECOND CLASS
RONALD JUDE AUBREY
UNITED STATES NAVY

for service as set forth in the following

CITATION:

"For professional achievement in the superior performance of his duties while serving as Senior Master Helmsman, Navigation Department, in USS AMERICA (CV 66) on 28 August 1986. Petty Officer Aubrey's alert response and execution of a series of rapid helm commands were instrumental in averting a potential collision with a group of poorly lit and illegally placed fishing vessels while the USS AMERICA transited the restrictive waters of the Straits of Gibraltar. His cool, flawless, and precise application of difficult rudder movements were indicative of his superior helmsman ability and avoided a possible catastrophe. Petty Officer Aubrey's superior performance, attention to duty, and professional ability reflect great credit upon himself and the United States Naval Service."

For the Secretary of the Navy,

R. C. ALLEN
Captain, U.S. Navy

LIBYAN LINE OF DEATH

I started out on the Flight Deck of the USS America launching warplanes. A dangerous job but you get some fresh air. On the down side, you work long hours. If it's hot outside, you're hot, if it's cold outside, you're cold. In addition, it is greasy, dirty, nasty work. I felt like I could provide more to the Navy than just being a wrench turner on a catapult. Not saying anything bad about the guys who do this work, my hat is off to them. I worked my way up to the bridge by studying for, and passing the test for Quarter Master third class. I had eight hours every other day all to myself. My first love was being on the helm. After a while, you can feel the ship through your feet. The ship talks to you, and you talk back by tweaking the rudder just a little, or a lot, left or right. You can be told this, but it's not real till you drive that way.

There are two modes of ship driving. One is Open Ocean with no one around. The second mode of ship driving is close in to another ship, or in a restricted narrow body of water. To drive under more exacting conditions requires months of experience and training. This is where the Master Helmsman comes in. As the senior Master Helmsman, I was the most experienced and the most nuanced in my driving skills.

Coming alongside another ship for refueling, or resupply, is a touchy business. Two ships, one hundred twenty feet apart, doesn't leave a lot of room for error. Once while alongside an oiler taking on fuel, we found an offshoot of the gulfstream, (a powerful ocean current) that had not yet been charted. We managed an emergency break away without damage or injuries. We spent most of the rest of the night crossing and crossing again to chart this dangerous current, which information we sent to the National Oceanic and Atmospheric Administration, (NOAA) for inclusion in warnings and updates.

So we found ourselves off the Libyan coastline some two weeks after the Libyan dictator had taken credit for a bombing at a club which was known to be frequented by American service members. Three aircraft carrier battle groups had assembled in the Mediterranean for an Alpha Strike on the Libyan coastal cities.

My watch station was Quarter Master of the Watch. I was responsible for tracking the ships position. Upon being relieved that night, I quickly went to the ships store. I was low on smokes, and then returned to the bridge in order to watch the ordnance loading and launching of the strike aircraft. My friend and fellow Master Helmsman, Karl, was on watch as Quarter Master of the Watch. As the first A-6 bomber launched Karl looked at me and asked, "Dude, why aren't you driving? This is history!"

I pointed to the young seaman on the helm and said, "You see that guy driving? It's his moment in the spot light. Years from now he can tell his grandkids he drove for the bombing of Benghazi. Maybe he won't care. Either way, who am I to take that away from him?"

Kevin thought deeply for a moment and said, "I didn't think of it that way."

It was afternoon when the word came down; an Iranian launching platform deck (LPD) had been caught laying mines just outside of the Bahrain channel. Our mission was to intercept the lifeboat that the crew had jumped into when the shooting started and turn them over to a U.S. Navy ship for transport to the Red Crescent. The Red Crescent is the Muslim equivalent of the Red

Cross. I was in the pilothouse navigating while the prisoner handling detail got briefed out on deck. The Commanding Officer of the detachment was Paul Evancoe, a Navy Seal; he was on our boat PB-758. Before we got to the life boat, Commander Evanco had us come alongside the Iran Ijar and he jumped over to the boarding ladder. We had no secure communications; we hadn't even been issued our crypto keys yet. But we did know there was a seal team onboard taking down the ship. It seemed incredibly dangerous to approach an armed and active seal team without prior warning, but that was the way the skipper lived—just grab it and go.

The prisoners we took turned out to be regular Navy, and did not seem to have an overwhelming desire to die for Allah. We would meet the Iranian Revolutionary Guard (IRG) later. The IRG is where they put the fanatics, and they were willing to die.

The Captain of the Iran Ajar turned out to be on our boat. He protested that we could not treat an officer of the Iranian Navy in such a fashion. The Boat Captain had someone put a sock in his mouth. No disrespect intended, just standard procedure.

AMBUSH AT MIDDLE SHOAL LIGHT

A patrol consists of at least two Patrol Boats. On top of that, you can throw in two more patrol boats, and maybe a couple of Special Warfare Insertion Crafts, (thirty-six foot low profile craft for operating in close to shore). In addition, we at times, had helicopter support. With this force, we could keep an eye on the entire area; utilizing intelligence assets, barge based radar, and of course, the various craft to conduct physical patrols of our area of operations.

It was a scene any military personnel would be familiar with. It was a "goat rope." The Lieutenants were yelling at each other, the Boat Captains were trying to figure out how to calm down the officers, and the helicopters were out flying the patrol route. All the confusion ended suddenly when gunfire erupted on the horizon, and the helos reported that they were returning fire. If you have never seen it, a 7.62 mm Gatling gun fired at night, looks like a laser shooting down from the sky. It also kicks your adrenaline into overdrive. Training kicked in, "Ronbo!" The Boat Captain called. Yes, that was my nickname, "Get me to that spot!" He was referring to the location of the fire fight. I could have interpreted the radio traffic, plotted the position, and laid out a course on the chart. I chose instead, to get the boat moving and drive to the spot on the horizon where I saw the light show. On deck, everyone geared up and manned their guns. The prisoner-handling detail was briefed and ready. The ambush had taken place at Middle Shoal Light, directly on our patrol route.

Two Boston Whalers and a Swedish made speed boat called a Boghammer had been sunk. Men were in the water; injuries and survivors were unknown. The men in the boats had seen the scout helicopter, but had failed to notice the two attack helicopters backing him up. When the Iranian patrol opened fire on the scout, he banked and flew in a direction away from the two attack helicopters, so as to draw fire. This allowed the two attack helos to do their job—still unseen by the enemy craft. The attack helos raked the three craft with 7.62 rounds and unleashed their two-and-a-half-inch rockets with flechet warheads. A flechet warhead has explosives in the center and packed around that is a mass of sharpened metal needles. They were very deadly within 15 meters. The enemy boats did not stand a chance.

We brought our boats close to the injured survivors, and one by one, had them swim over to the boat's stern. They were searched for weapons and tied and blindfolded before being laid on deck. The Boat Captain of the other PB had to disarm one of their prisoners. We were jealous; none of our prisoners were armed. Not all of the injured survived. One of their crew bled out from a head wound. There was not actually anything we could have done for him. First aid was rendered to all of the injured, but he needed a level one trauma center. He never made it to one.

The prisoners were transferred to the barge for processing and further transport. The Geneva Convention, the international body which determines the laws of civilized warfare, at that time, stated that you had to feed the prisoners the same thing the crew ate. Dinner that night was cold ham sandwiches. All of our meals to this point had been hot; we wondered what was up when the skipper walked in carrying a tray of sandwiches.

"Anybody want a ham sandwich? I offered them to the prisoners but I guess they weren't hungry." He had a very big smile on his face. Muslims, which most people in that part of the world are, cannot eat pork. The rules of warfare have been amended to state you have to feed prisoners in accordance with their cultural and religious rules.

During the helicopter transport of the prisoners, one of the prisoners either fell or jumped out of one of the helos resulting in his death. I never got that story straight-- but I was told he was not pushed.

It is amazing, how you go from total combat ready to afterwards—when it's all over and the adrenalin wears off. You eat; and then if possible, you sleep. We did our job. It was unpleasant—but necessary. The Iranians could have stayed home, but they chose to come out and play. I had a better day than the dead guy.

I finally gave him his own room in my head.

A good thing, too, since your dead are always with you.

CNN REPORTS

It was close to the end of our tour in the Arabian Gulf. We had recently had run-ins with a couple of helicopters, so we were getting itchy trigger fingers. The closer you get to going home, the more you feel like something is going to happen. This is in addition to feeling like something is going to happen as soon as you get on station. Not to mention, when you wake up in the morning, you know something might happen. It's kind of a stress-filled life.

Another day on patrol, I was navigating and watching the radar. Most of the crew was down below in the air conditioning, just trying to get out of the heat. It was only ninety-five degrees down below. My duties kept me topside most of the time. The only place worse than topside was the engine room. The heat from the engines kept that space a nice toasty one hundred-forty degrees year round. I felt sorry for the engineer; but I was proud of him. If he didn't have the part he needed, he just made it.

I was tracking an intermittent target on the radar. That happens sometimes when you're tracking an air contact with surface radar. I was scanning with the binoculars and saw a helicopter off the starboard bow. He was closing, and not answering on the air distress channel—which all aircraft are required to monitor. Sometimes, with non-U.S. aircraft, they ignore this rule. I took the boat to condition red, and our Electronics Technician ran up the ladder screaming, "We can't go to condition red, it's August!" He wanted to go home pretty badly.

I yelled, "Just track him with the twenty; do not fire!"

"It's August! It's August!" He yelled back.

Just after we started tracking him, the helo broke right and down, and quickly disappeared into the haze. We stood down from condition red and completed the patrol.

The XO was waiting for us when we tied up to the Barge. He informed us that the CO requested that we not target the CNN helicopter again. My answer was, that they should answer up on air distress as required. I asked if he wanted my log ...he didn't.

OIL TANKER ESCORT

Our official mission, was to help escort the reflagged Kuwait oil tankers—so somebody had to actually do an escort. I figured, we had paid our dues when we escorted the mine sweepers along the transit route. We were in front of the mine sweepers when I asked the Boat Captain, "If the mine sweepers are behind us, who is actually sweeping for mines here?" Some questions don't need to be answered.

The previous week, I had the picket duty. The command would trust this to me, but I couldn't take the boat on patrol. I was where they could keep an eye on me. No vessels were allowed to come any closer to the barge than so many thousands of yards. This is called a United States Naval Forces Security Zone. I stayed active on the radio but one ship didn't come up on the hailing and distress channel. We had to get his attention. So, I had one of the crew launch a 60 mm mortar flare. It turned out to be an oil tanker. A magnesium flair will burn through a ship's deck in about ten seconds. We got his attention; just missed the ship, too.

It was a busy week: We also intercepted a Russian warship. The Russians don't think much of being told to change course; but they got the standard speech about under water hazards that everyone else got. I thought it prudent to have the large gun forward turned to starboard—in order to prevent any misunderstandings. I most definitely did not want to start a war. Especially I did not want to die first, that would be a very short war.

We joined our escort convoy as they went by the Barge Hercules at ten that morning. The convoy was doing about 12 knots. Boredom set in quickly. We were under the tactical control of the cruiser leading the whole show. He put us a half mile in front; I called that, "Just checking for mines."

When our Boat Captain couldn't take the boredom anymore, he told me to get him if anything came up; he was going aft to sunbathe. In the pilothouse with me was the helmsman doubling as throttle man. In action, both of these positions would be filled. I stayed busy navigating and working the radio. The cruiser said there would be a flyby of one of our helos in a few minutes. I stepped out on deck as the helo went by. In the 40 mm gun mount one of the crew had put a mannequin. On the port side gun mount, someone had tied the mannequin to the twenty mm anti-aircraft gun. On the deck aft was the Boat Captain sunbathing. The door gunner for the helo was leaning way out in his harness, and in his hand, he had a video camera. I shook my head and walked back into the pilothouse. Amazingly enough, we never heard about the video footage of this unusual procedure for manning the boats weapons. Like everything else, I am sure it is on the web somewhere.

Upon reaching the Port of Kuwaiti, we were released by the cruiser to anchor overnight. We would pick up a south bound convoy in the morning. I selected a nice spot about a thousand yards off of an island just off the coast. We set the anchor, and the boat captain, along with the rest of the crew, decided to go bug hunting. Bug is a nickname for crabs. Particularly tasty on an escort operation.

I had just been transferred to this boat after the boat captain of my last boat had gotten himself fired—I forget the reason. My new Boat Captain was Vito—that would be short for Vitigliano. Tall, handsome, with classic

Italian good looks, he also had a tendency to play a bit. And this play usually involved some antics that the CO found objectionable. The radio frequencies were unusually quiet, as I kept most of my attention on the radar. Just as the crew were all suited up to go hunting for crab for dinner, a small skiff pulled alongside. I believe the uniformed gentlemen were Kuwati border guards. They carried 38 caliber revolvers. Our boat's gunner pulled the retaining pin on the fifty cal mount, I smacked a magazine in an M-16 and stayed just behind and off to one side of the Boat Captain. Vito chose to remain unarmed.

No one pointed any guns at anyone; it was just the power of presence. The border guards saw our weapons. We had them so seriously outgunned, we just smiled. Vito chatted for a few minutes, and the Kuwatis' smiled and went on their way. I went back to watching the radar and checking the radios. We were supposed to call in every hour. I noted the absence of any answer to my radio calls in the log, and kept going about my business. Loosing radio contact during an operation is a major issue. Loosing radio contact when you're at anchor, it's kind of expected.

A couple of hours later, with all hands back on board, and crabs being handled by the cook, we prepared to get underway. The border guys showed back up and said, "You go now." We smiled and waived as we got the anchor back on deck and headed a little South by Southeast making about seven knots. I logged everything. I always logged everything. That was my job.

About two miles out from our anchorage, all of the radios came to life at once. Everyone, and their mother, had been trying to reach us. Channel 16, the hailing and distress channel had a cruiser on it. The operation frequency had the barge on it, and we heard our immediate boss on the air distress frequency. I looked at Vito because I really didn't know which one to answer first. Vito grabbed the air distress mike. I figure he wanted to talk to a familiar voice.

We were ordered to rendezvous with the cruiser and report our heading speed and position every fifteen minutes. More interestingly, we were asked, if there were any friendly killed in action (KIA), or missing in action (MIA). "What action?" We asked ourselves. We had just been bug hunting.

A ship is a great place to start a rumor; you just need a few ingredients. A boat out of radio contact, is not unexpected, but could be trouble. An inquiry by the border guys had gone up their diplomatic channels. A ship's Captain had been asked by an Admiral and so: This is the best story that I heard, and this was after we got back to the barge.

The fact is, that the island we were anchored near had a radio station, and we were in their shadow. They could hear us, but nobody else could. We couldn't hear anybody until we moved. The rumor said: The guys had gone ashore and the radio station said they were under attack by U.S. forces. It gets better! There was gun fire! Vito had been taken prisoner! We had attacked the jail where he was being kept! The U.S. was on the verge of war with the Iraqis'!!!   All we did was go bug hunting.

We weren't even tied up to the barge before the CO, red-faced and waving his fist in the air, was screaming for Vito. Vito turned to me and said he wanted the log. We had already reviewed my most excellent log on the way to the barge, and determined, we were blameless. Vito went to convince the skipper. I went below to see what the cook had done with the crab. That wasn't the only time my log saved Vito.

CAMELS ON THE LOOSE

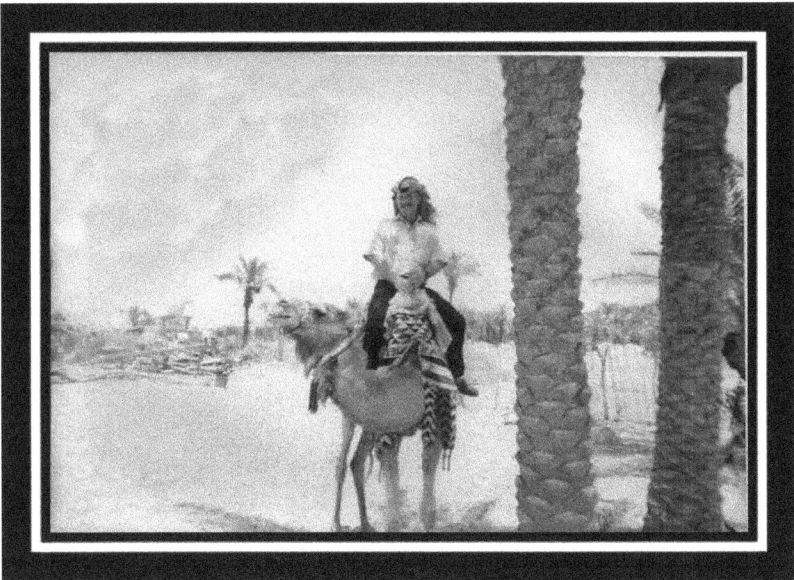

Aqaba, Jordan is a lovely little seaside Arabic town. My ship made port call there. Arrangements were made for beer on the pier. The beer was cold, it was cheap, and it was American. Some of the crew never got off the pier; but as for myself, I had to check out the three first-class hotels, just off the pier. Their beer was cold, American, and very expensive. The hotels won because I couldn't see the ship from the bar.

My buddy, Adam, a corporal in the Marine Corps, found the full size piano in the hotel bar; and he is actually a very good pianist. That was all good. He got to entertain, and I got to drink. Then the night came, when I just happened to be on Shore Patrol, and all three ship's captains, the Admiral in charge of the group, and the military governor of the area, came in for drinks. This is military to military politics. The regular piano player was there, so Adam started singing—which he does very well, also. The Shore Patrol kept the worst of the American drunks away from the V.I.P.s. It was too important a meeting to let anyone get stupid in public. I have seen sailors get very stupid in public, numerous times. We did manage to make it a quiet night at that hotel bar, even if Adam did flirt with the ladies in the audience a little more than I was comfortable with. Why was I uncomfortable? In Arabic countries, they cut off a hand for thieving. I don't want to know what they cut off if you steal someone's lady.

I continued my exploration of the town. Across from the hotels is an open lot A tribe of Bedouins had set up camp and were giving camel rides to the tourists.

I thought it would be a unique experience, so I went for it. Camels are nasty creatures; they smell bad and have a really poor attitude. The camel I chose was all about being cool with me getting onboard, just like mounting a horse. I even got to wear a Keffiyeh, which is the traditional Arab male headdress.

The camel bends its legs one at a time to get down for boarding, and then it stands up one leg at a time; it's a rough process. Then before I was actually ready, the camel herder got him going. Left, right, left, right, now I was rocking from side to side. I tried several different positions and finally found one that worked about half way. I had reins for all the good they did. I never got the camel to turn in the direction I wanted him to go, but when I pulled his chin into his throat, he did stop. He looked back at me with pretty much blatant hatred. I knew just how he felt.

I was on my way back to the ship after shopping in the bazaar, when I looked out the window of the bus and saw the Bedouins chasing the camels around, and

139

whipping the heck out of them with a long stick. The camels probably deserved it. I later heard it took till morning for the Bedouins to round up all their camels. I did get a cool picture out of the whole deal, which my sister Vickie has on her refrigerator to this day.

OPERATIONS OFF THE SOMALI COAST

I saw fireworks and thought, "The last time I saw something like that, was off the coast of Somalia." For just a minute, I was there again, driving a landing craft off a hostile coast. I was in my element and quite happy.

The U.N. peace-keepers needed to evacuate from Mogadishu, Somalia. The U.N. had given up on that Godforsaken part of the world. The Marines went ashore to cover the peace-keeper's withdrawal. The Marines could cover their own withdrawal. We had everything for that op: Amphibious Assault Vehicles, Landing craft, attack helicopters and an Amphibious Ready Group to back it up.

My claim to fame was that the landing craft utility's (LCU) Quarter Master had broken his leg and they needed a replacement. My Chief, not much of a combat type, but he was great with the paper work, made it very clear that I didn't have to go. In my mind, I had small boat and weapons experience. I had an obligation to go. There was a bit of conflict between the LCU chief and my shipboard chief. The LCU chief finally told my chief, that I was with the LCU crew getting ready to go into harm's way, and that my chief should go back to the bridge and do some paper work—which was all he was good for. In the world of men, that is one heck of an insult.

Amphibious craft, full of Marines, hit the beach at the airport. This would be the security detail and last to leave. Marines then fanned out to the port and strategic positions to cover the retreating peace-keepers.

The LCU's crew and I spent the day readying weapons, loading magazines, and checking for proper operation. I pulled the speed loader out of the belt of M-16 ammo and fitted it to an empty magazine. M-16 ammo comes five rounds to a clip, the clip inserts into the speed loader. Then you just push all five rounds into the magazine at once. The LCU's gunner looked at me and said in awe, "What did you just do?" I explained the mechanics of the operation to him, thinking in my mind that he was testing me. It turned out, that I had more small arms experience that he did. After that, he just called me shooter. I have a coffee mug with that name on it. I consider it quite the compliment.

We pulled into Mogadishu harbor to load up with our Marines. After several hours, and one sniper alert, yes I grabbed my M-16, we took the Marines back to the ship and got underway for further taskings.

The General in charge of the whole shooting match had been adamant about being in the last amphibious vehicle to leave the beach. When the general wants something, he gets it. Unfortunately for him, his amphibious craft took on water and sank. The General and crew got off the craft. All of this took place against the background of tracer fire going into and coming from the beach. At this point, we were out of range of the small arms the insurgents had. We found the General and crew and took them onboard. He did a very meaningful thing.

The first words out of his mouth were, "Do you have someplace my crew can get warm?" When the General asks, you answer quickly. We put the crew below decks in the galley. We gave them our extra blankets and all the coffee they could handle. That impressed me greatly. A General should take care of the men under him, but to see it happen, is beautiful.

The second thing the General asked was, if we had secure communications. I stepped out of the way so the General could get to the radios. For a few precious minutes, we were the General's flag ship; which was pretty cool. Then, we made the well deck of the amphibious ship we were attached to, and the General and his crew did a MacArthur and walked through the knee-high water to go back to work. We got back underway, and followed the battle group up the coastline.

SOMETHING IN THE WATER

Just another boring mid-afternoon on the Barge Hercules; do a little maintenance, throw a little football, duck as the two fighters blow by overhead at low altitude at high speed. The lead aircraft punched off his spare fuel tank to grab more speed.

What the heck!!

Iran and Iraq were at war: An Iranian fighter was chasing an Iraqi who had wandered into the wrong airspace. The barge had been at the same spot for so long, the Iraqi's had started using us as a turning point. The Barge was moved shortly after this and the Iraqis bombed their own Island by mistake.

We ate dinner early because the skipper wanted us to recover the fuel tank that the Iraqi had dropped. The Sun was low on the horizon when we got underway. We approached slowly and Vito called "All stop!" I was confused. Hooks were on deck attached to lines, everything was ready. Vito said, "I see something in the water near the fuel tank."

I looked with the binoculars, "I got nothing." I reported.

Vito grabbed my arm and said, "Look closer. There is something in the water, and it looks just like a mine."

I smiled and said "Logging it!"

I knew the game now. Proper procedure for a mine at that time was to blow it out of the water. That was later changed to send EOD in with explosives to detonate it properly. It seemed that some of the mines fired on simply sank a few feet and exploded later. We backed off to six hundred meters—that was right out of the play book. I was on the radio and got permission to open fire on the "mine like object" The sun had set; we needed light.

"Lume!" Vito called. The guys fired the sixty mm mortar right up.

We had illumination flares hanging from their parachutes. Heck, I even fired a few from my M-79. The MK=19 grenade launcher and a couple of fifty cal's got in on the act. I even saw somebody grab an M-60 and start popping off bursts. It was great; we relieved a lot of stress. Finally, we reported area clear. The Barge inquired as to whether we had retrieved the fuel tank. We reported that it had taken collateral damage and was sunk.

All was nice and calm until we got alongside the Barge. We heard a familiar voice bellow, "Vitigliano!"

Vito quickly said, "Wheels. You and your log, with me."

I got to wait in the passageway outside the skipper's stateroom, while Vito went inside with the log. It was pretty muffled, but just at the end, I did hear the skipper yell, "Get out!"

Vito came out smiling.

It was a good day.

OUT ON A LIMB

Devin, the children and I went to Otter Creek Park for the afternoon. Just a little hiking, that's all we had planned.

The simple plan changed when Josh and I saw the almost vertical mud covered cliff face. We had to climb. He was wearing tennis shoes; I was in my western boots. Western boots are made with a smooth sole. Not the best climbing gear, but I was confident I could handle it.

I almost did too.

Three quarters up Joshua was in front, I was coming up close behind him. I saw a perfectly good root to grab onto, but when it pulled away from the cliff face, I saw it was just a stick stuck in the mud. Quite involuntarily, I began a slow slide down the cliff face. Unable to find good footing, I tried to slide a little sideways so that I could grab a small tree I had seen on the way up. The small tree went between my legs and I grabbed hold and wound up upside down. My son was jumping like a mountain goat down the cliff face that I was unable to find a foot hold or hand hold on. He stopped and looked at me, "Father?"

I answered back, "I'm good."

I decided to hang out there while Josh continued up to the top.

Afterwards, we walked up to Devin and asked if she got the picture. She replied, "No."

We said in unison, "Now we got to do it again." We turned and took one step. This was when we felt the full force of Devin's glare. Some women send out a type of psychic energy when they get angry. The force that hit Josh and I stopped us in our tracks.

Without turning Josh said, "She's looking at us, isn't she?"

I answered, "Yes."

Josh said, "We should go back, shouldn't we?"

I answered, "Yes, we should."

This is a force that you have to feel to believe.

There is more to this universe than science can explain.

Chapter 6
Paul Benjamin Earl Merk
US Army
SOLDIER UP!

If you are headed for Army boot camp, I strongly advise you not to go in the summertime. For one thing, it's hot, and for another you could die, like my nephew almost did.

This is Ben's story:

Soldier Up!

(Written by Ben Merk)

Well, let's see if I can remember the details...

I shipped out for Ft. Sill, Oklahoma just a couple months after my 20th birthday. I can't remember my exact MOS, but I think it was 25 Sierra. I had wanted to be in satellite communications, and thought, with an 85 or 87, whichever I got on the ASVAB, but alas, my engineering score on the test was just under par. Or so my recruiter told me. 25 Sierra was the next best thing; setting up satellite relay stations. Seemed like an exciting enough career, and was linked with SatCom, so I was sold on it. I think I shipped out around June 20th. I was headed to the armpit of hell for the summer.

My drill sergeants were Drill SGT Kempen, and Drill SGT Mills. My head Drill SGT was Drill SGT Kempen; he was about my height and Caucasian. Despite his physical size, his intensity always made him seem more imposing. He took nothing; and smoked us more than any drill sergeant smoked the other platoons, hands down. We were the 4th platoon Young Guns of the Delta Dogs Battery. And we were also the only platoon to get smoked the morning of BRM qualifications at the range in front of everyone else. Drill SGT Kempen definitely kept us humble and tired. Especially, considering the fact that if we tee'd him off on bivouac, he had no problem putting on on 50-50 fireguard rotations. He was perennially tee'd about having been pulled from teaching sniper school in Alaska, to teach a bunch of newbies how to stay alive.

Drill SGT Mills, on the other hand, was a much more physically-imposing man, but had a surprisingly gentle demeanor—until pushed. He was the good cop to Drill SGT Kempens bad cop. Standing about 6"2, or 6"3, and weighing in at what I would have had to figure at 240-260 pounds, of mean-looking black man. But, like I said, he was the nicer of the two; looks can be deceiving.

Both were excellent at what they did. I distinctly believe they held us to a higher standard than any of the other platoons in our battery.

As for memorable individuals, I'll start with the best and work down to the worst. If I'm starting with the best, I better reference my battle buddy, Trosclare. I remember little about him other than his goofy big-eared appearance—and his constant hilarity. He was lined up in the bunk across from me and that made him my battle buddy, according to the drill sergeants. He was the guy who had a lot of heart and a really hard head. He had no problem embarrassing himself, or in some cases, others, but mostly himself. He was never in peak fitness, but he, like me, tried to stay in the very middle of the pack; especially when it came to attracting the drill sergeant's attention. When it came to getting the platoon's attention, though, that was his specialty. He was the guy who took the $15.00 bet to smear icy-hot all over his privates. And that was just the kind of stuff he'd do; whether to get a laugh, get a buck, or just to get attention.

Then there was Baley. He was probably in his early 30's; had a wife and a couple of kids at home. He was a little more timid, but took things very seriously. Probably as he should have being the family-man he was. He was never the best, but he had a certain quality that made you begrudgingly accept that he was somehow in charge. It was that leadership-by-default thing, I think. By filling that role naturally, he stood out to the drill sergeants. Since he seemed to want to take the weight, they did make it official for some time, giving him temporary titile of platoon leader. He got some of the credit for our good behavior and caught the brunt of it when we were out of line. I hate to say it, but there were probably more than one occasion he rubbed the group the wrong way, and our behavior sank simply to see him forced to weather the storm of responsibility. Had we known who would eventually replace him as platoon leader, we probably would have acted right for him.

Which leads me to Private Bush. He was the worst of the best. An Arena League football player in his mid-20's, 26 if I remember correctly. He was chunky around the middle, but he was conditioned to keep up with just about any physical activity we were presented with. He was the jerk jock that most people hated in high school. He was more than happy to proclaim his greatness and his right to leadership, but very infrequently helped those in the platoon who needed a hand up. He was more of a push-you-down, kind of guy. Fourth Platoon was also the only platoon who, after a certain point in basic training, weren't allowed to have any of the sweets from the mess hall.

For most platoons, that started between two to four, needless to say, our drill sergeants were the no-frills kind of guys, so that wasn't the case for us. Eating any of the good stuff equaled one serious post-meal smoking. I mention that to reference that Bush was also the guy who ate three cakes every day after week three, for the simple sake of being a self-serving person. On another note, he had the highest pitched voice and was absolutely hilarious to listen to him holler, "Yes, Drill sergeant!" It was a tiny, little chipmunk voice. It was almost impossible to take him seriously. He also had a short-run as platoon leader, but his ego and arrogance lost him the position rather quickly.

There was a really good guy whose first name was Mark. I think he was the only person whose first name I actually knew. He didn't talk much; he was strong and had been a golden-gloves boxer in New York before joining the army. He did his thing and mostly kept to himself. We had a few conversations; some funny, some serious, but always good. I respected the fact that he could have bested almost anyone there, but he kept his ego in check. He took no stuff from anyone, and dealt none to anyone if he could help it. He also helped to encourage our weaker men in the platoon. He was just a really standup kind of guy. I was 20, and he was about 28. And he was the one guy I looked up to in the platoon. He had a short run as platoon leader, but saw it end quickly because he never had the desire to be the center of attention. No fault that I remember, simply no desire to be in charge.

Now we start the worst list: Of course all names are changed to protect everybody involved.

There was one, fat, egotistical, self-centered, brain-dead, ugly rat-of-a-man, who had come into training with a civilian nick-name of "Bull". This guy was a less-talented, less-gifted, less-capable version of Bush. Those two actually got into a fist fight once. My theory was there could only be one ignorant, self-absorbed, egotistical, incompetent rat-of-a-man at a time, trying to claim the reigns of leadership. Well, this same punching bag started to pick a fight with me. Or perhaps, I started to pick a fight with his whiny-loser self one night when we were making it back late from a bivouac.

We all had sand in our butts, and were pretty chaffed and tired. I remember the situation. He was griping and picking on one of our weaker privates, as he regularly did, being the bully that he was. It's always been my nature to stand up for those who have been picked on, so I started to insert myself into the confrontation. He wasn't exactly thrilled with that because I wasn't the type of guy you could bully. It led to a bit of a standoff. Finally, after a bit of verbal prodding, he pushed me.

I was about to raise him off the pavement when the aforementioned Mark, grabbed me from behind (I didn't know who had grabbed me at the time). As I was grabbed from behind, the spineless rat-of-a-man, sucker punched me in the face; and then stepped away as several people wrestled me to the ground. I fought until I found out it was Mark who had grabbed me.

After I calmed down, he explained that he grabbed me to keep me from injuring Bull; and in the process, hurting myself—and the platoon. Mark was wise, and I think he knew what I was capable of; which at the time, I truly didn't know myself. At the end of the day, Bull got the sucker punch, but I kept blood off my hands.

Let's see, our physically weakest private was Private Bells. I always felt so bad for him. He was just a weird guy who had no strength coming into bootcamp. I remember seeing him at the unit where they prepped us before sending us on to real basic training. He couldn't seem to stand still. They were always making him do pushups; which he wasn't capable of doing. And he was always misplacing his Army handbook; which was supposed to be on us at all times—which led to more pushups that he was incapable of doing. He wasn't a bad guy. He was just ill-prepared for the task at hand. He was the guy that bully-types flocked to; and the guy I always felt compelled to defend. He taught me a lesson though.

One time, the final time I tried to defend him, he actually turned on me in anger, because he had decided he wanted to fight his own battles—even if he knew he was going to lose them. He may have been the weirdest and the physically weakest, but whether he had it when he started, or developed it while he was there, he had strength of character. After that one instance, I respected him and let him alone to fight his own battles.

Last, but not least of our very worst, was the private I only remember as Craige. He was that guy that immediately struck a nerve with the drill sergeants; and he kept on their worst side throughout. He didn't do himself any favors by not shaving on a daily basis. He got smoked and pointed out more than anyone else, but mostly with good cause. I think it was two weeks into training before they took his rank. I think he was somehow coming in as PFC, and by the end of training, he was forced to start at the beginning of training with a rank of E1. He was the guy who would sneak down to the PX and purchase cigarettes, and dip logs to sell for quadruple the price to the other platoons. When Drill SGT Kempen found out, we all wanted to beat him to death; because we got smoked from midnight till 3:30 a.m. He was just a waste of space while there.

That's pretty much it for all the characters I found memorable enough to be of mention.

As for me, I did my best to stay really middle of the pack. Although I was always the loudest when I came to yelling "Sir, Yes, Sir!" I guess despite everything I did manage to stand out, some. I was notorious for working with our weakest links to help pull them up. Some weak links liked being weak, and others genuinely appreciated the support.

I remember on our 5k march out to the BRM range, being at the very back of the pack in formation with Private Westley. He was another guy who just didn't have a lot of physical conditioning. And other than this instance, wasn't profoundly memorable.

Just behind us, were the 1st Lt. and a cadet recently out of West Point there for training. Behind them, was a transport vehicle; which I eventually figured out, was there for picking up all the privates who fell out on the march. It wasn't half-way into the trip when others started lagging behing. Westley included. Every time anyone would fall behing, I would run back to give words of encouragement; or tell them to hold onto my gear and I'd help drag them back to the pack. My being vocal about it, did lead to some of my other platoon mates to insist that I shut up. But I explained that no one in Delta Dogs battery should fall out on a 5k. And at the very least, 4th platoon Young Guns was gonna finish with every man that started.

They still weren't happy about it but I never shut up. And neither the CO, nor cadet ever said a word to me about shutting my mouth. I figured I was in good shape. Needless to say, you can't save everyone. People I tried to drag along fell out. But Private Westley dug deep within himself and found the courage and the will to continue.

4th Platoon Young Guns was the only platoon to finish with the same number as we started with; much to our drill sargeant's surprise. Weastley tried to thank me, but my only response was that I was simply the motivation and he was the motion. He had done all the hard work, and was much stronger than he had ever known. That was a really good day, and we got to start playing with M-16's, which made it even better.

There were a lot of fun times in basic training. There was our first big group activity; an obstacle course. Not everyone completed it; most caught up and fell during the rope crawl at the start. I remember being so nervous for this thing. Heights have never been my favorite by any means. I remember having difficulty, but making it through each and every obstacle until I finally reached the repelling station. Man, my confidence had been climbing and climbing with each obstacle, until, of course, I looked down that wall....

Well, drill sergeants have a great way of getting you motivated to do things you might otherwise avoid. Drill SGT Kempen said something like, "Merk, if you don't go now, I'm cutting this friggin rope!" –while I was hanging straight out over the ground. Yes. That got me moving with quickness.

I remember Drill SGT Mills passing out mail and laughing as he cracked jokes on different people; and how everyone couldn't wait to do the 25 pushups required per piece of mail you received. I also remember the deep sorrow when you seemed to go weeks without seeing a single bit of news from family or friends.

The gas chamber was one of those experiences you have to love the build-up for. I remember them having us all seated on the bleachers giving us all the warnings. They pulled an awesome joke that is probably done at every boot camp. It went something like this:

"Now privates, close your eyes because this next question is very important." He made an off-color remark that made everyone laugh, and broke tension. Everyone cracked up, and then the drill sergeants quieted us down, and he got really, really serious again.

"It is important you do well here, or this gas chamber could cause serious injury, or in some cases, even fatal."

I remember the burn of trying to scream "I want to be an Airborn Ranger, I want to lead a life of danger!" – ten times before our drill sergeant would let us try to start out of the room. I say try because as we were walking down the side wall, two drill sergeants kept jamming us against the walls and holding us back from getting to the door.

At about that point, I thought my throat was closing up from the gas, and I was so happy when we finally got outside. But my breathing didn't get any better in the seconds after I hit the door. It felt like I was being asphyxiated. I saw a drill sergeant and tried to explain I couldn't breathe at all. He started to examine me and then cracked up laughing as he pried the hands of the battle buddy behind me, off my collar, which he had grabbed instead of my shoulder. He had twisted it so tightly, I was actually being asphyxiated. Once he let go, oxygen came back; and the only pain I had left was my eyes, which stayed that way until long into the night.

I was there through the fourth of July. Because of that, we got a unique privilege not many others got; a chance to be a part of the crowd. Montgomery Gentry was playing that night on base. Those of us who were willing to lose the sleep, were allowed to take a long march to the concert. We ate and drank as we wished; as much soda and pizza as we could afford. I went, enjoyed myself, and had a mighty good time.

What most of us didn't think about, was that the next day was our Pre-PT test, to evaluate where we were at about week six. When we got in that night, rather late, they smoked us out in the commom area to the viewing pleasure of all the smart privates who'd stayed in that evening to be properly prepared for their PT tests.

I'm not gonna lie, it was worth every painful minute of that punishment to have that short period of freedom again. The next morning, I got up and kicked butt. I didn't score a 300 or anything, but I was pretty close to my personal best at pushups and situps; despite the late night and two tons of junk food I had eaten the night before. What I was really excited about was, that I did a two mile run in 14:28. A personal best for single mile time, let alone to be able to have paced a 7:14 mile for two in a row! I was proud and rather exicted. Little did I know, that around a week and a half later, I would be finding out—I had a bad heart.

It was towards the end of basic training, probably week seven--or the middle of week seven—when we were coming back from a three-day bivouac. I was worn down, exhausted, my body was probably still recovering from a really bad case of laryngitis I had a few weeks before. Needless to say, I had been ground into dust at that point.

That night, we didn't get much sleep, or maybe it was just me. I don't remember getting smoked or having any problems, but I remember wearily pulling my fireguard shift and praying for my last hour of sleep to be very restful and regenerative.

That morning, we woke up for regular PT, and our ability group runs. I started the run and felt funny but kept on going.

On the return portion of the run, I thought it was funny that my left side was going numb. I was having real difficulty keeping my arm pumping in rhythm. Eventually I slowed down.

It's funny, but I kind of laughed to myself, about how I was probably having a heart attack. The idea of it seemed so absurd, considering everything I had done physically in training.

Well, it was the only run I ever fell out of and I think I had set an expectation for myself. I half-hobbled, half-jogged into the finish. The captain called me out for having fallen out. I tried to explain I had gone numb, and was still numb and having trouble breathing.

They weren't buying it and said, "Sprints, NOW!"

I'm kinda hard headed, so I started hobble-sprinting with my left arm limply flopping around back and forth until they said they's seen enough. Not long after that I explained to my drill sergeant what was going on, and they put me on a special solo trip to the medical center.

When I was finally seen, my blood pressure was still ridiculously high. I didn't know that at the time, because I had no idea what my blood pressure was supposed to be. I don't remember the exact numbers, but Mama could give them to you. I'm pretty sure she remembers.

(220/120, heart rate 220—courtesy of Ben's Mama, typist).

That led to several days of continued testing at the actual hospital facility. I wore a heart monitor and some other fun stuff like that.

It was like three days after the initial heart trouble, we were waiting in lines to get our dress uniforms. I was going through line wondering if there was any point in my waiting in line at all. Apparently, Drill SGT Kempen was thinking the same thing. He was at the very end station and asked me why I looked so sad, and if I had been wasting his time.

I looked at him, and in my sadness, simply said what my blood pressure had been, and that I didn't know what it even meant. He pressed on and asked me if that was bad. I responded that I didn't know, but the doctors seemed to think it was. Apparently, he asked someone what that meant because he didn't say another thing to criticize me.

The last time I saw him, was just before I was going to be sent to the injured unit. I waited outside the office, and when he walked out, I told him that I was sorry for having wasted his time.

It meant the world to me when he looked me in the eyes and said, "You didn't waste my time, private." He turned and walked away, and I went to begin packing my belongings; feeling just a little better about having failed everyone in the world.

The last time I saw Drill SGT Mills, was when he was transferring me to the injured unit itself. He took me up to the closet where we put all of our civilian belongings, eight weeks earlier. As we made the short journey, I explained that I was sorry for failing. But, that I was going to go home and get my health sorted out, go to college for a few years, and come back as an E4 specialist. I think he appreciated my enthusiasm because he smiled as I spoke about my desire to come back.

It was at that time he said something I never expected anyone to say to me, let alone an enlisted officer.

"Son, you did a good job. And if you ever come back, I strongly suggest you apply for officer candidacy school."

I was more than a little shocked, and even thinking about it now, I still can't believe he thought I had leadership quality. Those two small comments from both of my drill sergeants helped soften the blow of failure.

It has taken years, and lots of love from family and friends to really dull the pain of bottoming out of basic training. But I know now, that God has had a plan all along. And now I'm working to bring those same skills to light in myself to brighten the lives of others.

I will never forget the things I learned, and the experiences I had in training. There are still times in my life today, when I am in the right mental condition, I tell myself to "Soldier Up!" – to accomplish things I have trouble believing I'm capable of.

Uncle Ron, thank you for having me write my story. I'm sure this is more than you were hoping for or expecting. But I wanted to share the mostly whole of my memorable experience at basic training. Feel free to use and discard as much as you see fit. This was the first time since training, I've put these thoughts and emotions down on paper. It was nice to pull them out of my brain and give them a tangible space in reality.

I love you very much and I appreciate your support for me after basic training. I'm proud of you for writing this book. I am much honored to think that you would want to put my story in as well. Sorry it took me so long to get it to you. And it was a little longer than I thought it was going to be. I look forward to seeing you again, soon.

Sincerely,

Your nephew,

Paul Benjamin Earl Merk

Chapter 7. Joshua Aubrey,

U.S. Marine Corps.

"For over 221 years our Corps has done two things for this great Nation. We make Marines, and we win battles."

General Charles C. Krulak, USMC (CMC)
5 May 1997

Joshua Daniel was born in Portsmith, Virginia.

So was his sister. The day she was born, he picked her up and held her in his arms. After all, no one had told him he couldn't. He has always been a very protective son; whether it's his sister, or his family, or his country. And he still is.

CARTOONS IN REAL LIFE

As a young boy, Joshua, as all American children, watched a lot of cartoons.... maybe a few too many. We would go to yard sales on Saturday mornings looking for any cartoon videos we could find. I thought it was a fun type of treasure hunt that kids would enjoy.

It was just another sunny Saturday day in California. I came home from work to a yard that was short a tree. The following story was told to me.

Joshua was bored, so he climbed the tree out in front of the house. Being a young boy, about nine years old, he enjoyed the view from being up a little higher than an adult's head. He didn't get to enjoy that view for long. As he went to climb up a little higher on the tree that had grown at an angle, he heard a crack. If you have ever heard the crack of a tree as it starts to fall from too much weight, you know it is a strangely, distinctive sound.

As the tree slowly began to keel over, Joshua came up with a plan. At the last second just before the tree hit the ground, he would jump, much as he had seen his cartoon characters do many times. His timing was just a little off. When the tree hit the ground, he bounced off of one of the tree limbs. He ran inside screaming, "Mom! I was just in a cartoon! The tree fell down while I was in it."

Years later, hoping to give Joshua's spirits a boost, I mailed him a collection of Warner Brother's cartoons. Sure enough, the next Saturday morning he got up early and went to the squad bay with his DVD and small boxes of cereal. This is in a Marine Corps Barracks where all the men are super manly. Someone came into the bay and said, "What the heck!"

You really don't expect to see cartoons in the squad bay. Soon, he had half a dozen Marines watching cartoons and eating cereal. The watch was making his rounds, and when he came into the bay, he too asked what this was. Joshua replied, "Saturday morning cartoons. Want some cereal?"

The Marine on watch stayed a few minutes to enjoy this unique brand of entertainment.

STORIES FROM HIS FATHER

When my son Joshua, left for Parris Island, we anxiously awaited for the first piece of mail from him so that we could begin sending letters. It is part of the Boot Camp program, that as soon as you class up, you are instructed to write your family and let them know you arrived safely. My experience at Great Lakes, Navy Boot Camp was a little odd. One of my fellow recruits had left home as a result of some problem and no one knew where he was. He had a private chat with the Company Commander, and was assigned other duties while the rest of us wrote home. Joshua's letter arrived about a week later. We were pleased to learn that Devin and I could send a small picture of the two of us. We quickly got that in the mail.

During Joshua's time at Parris Island, we had a little weather problem (first hurricane to visit Louisville,

Kentucky, in fact, the region.) A wind storm had taken out the entire citie's electricity; and a month later, an ice storm—which did the same thing all over again. If everything else fails, the mail will get through. I had to hand write some, but kept the mail going.

Getting mail is a very big deal in Boot Camp. It helps to keep your spirits up, and reminds you that someone in the world still loves you. Along with each letter I wrote to Josh, I included a short story or poem with a military theme. He shared these writings with some of his fellow recruits; whom I had a chance to meet on graduation day. We had gone to the barracks to pick up Joshua's sea bag and hanging clothes bag, when I heard one of them ask, "Is this the story guy?" I was in shock that someone would refer to me in such a way. But also deeply gratified that I had made such an impression on people I didn't even know.

Josh said yes and introduced us. The two young freshly minted Marines, who happened to be African Americans, wanted to shake my hand and tell me how much they had enjoyed reading my writings. It seemed they approached Josh one day, had heard of the stories and poems, and wanted to read them. In boot camp, anything from the outside is a God send. I thanked them for their service. Devin and Ceara were with us at the time, and they were very impressed with the young Marine's manners.

Following Josh's successful graduation from Marine Corps boot camp, the prayer above was enscribed.

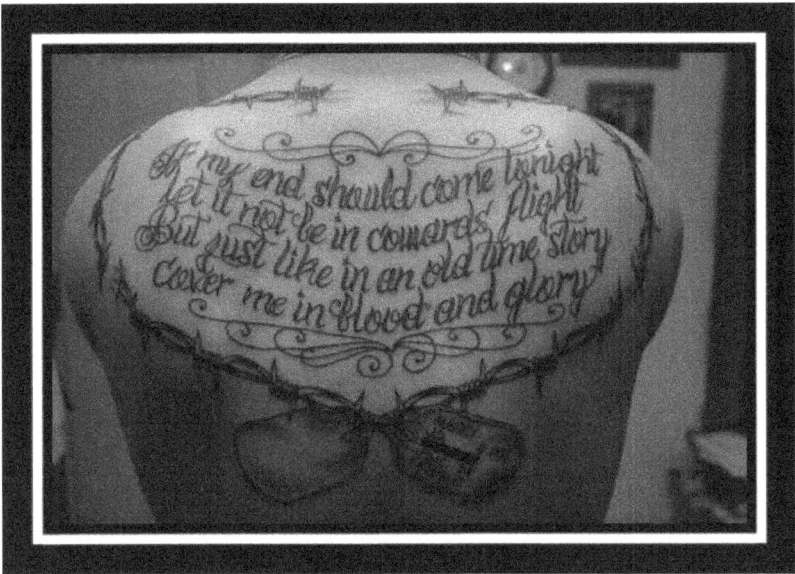

If my end should come tonight
Let it not be in coward's flight
But just like in an old time story
Cover me in blood and glory

HUH RAH!

Chapter 8. Bootcamp Stories

INTRODUCTION

The following stories are a compilation of short essays drafted and sent to Joshua Aubrey during his time in Marine Corps boot camp on Paris Island.

These stories were symbolic to help improve his morale and spiritual strength—coming through as a fifth generation warrior.

Understand your mission:

Don't die for your country-
Make the other poor bastard die for his!

DUTY

He asked no questions, named no price.

He sought no reward.

Women, children at his back.

He placed one hand upon his sword.

Enemy at the gates.

He took his place.

He thought of home.

He thought of love.

HONORS

The day was bright and sunny; just warm enough for a parade. The marching units were well separated in distance to give the spectators a chance to shift focus from one to the next. It was a good day.

Along the route, an old man sat in a wheel chair; his family at his side and behind him. The doctors had said he had less than a year to live. His response was, "I've had a damn good run." He had insisted on going to the parade.

As the Marine unit approached, he handed the blanket which had been covering his legs to his great-grandson. Then he took his old campaign hat from the bag on his wheel chair, and placed it squarely on his old, balding head. Then the old man stepped out of his wheel chair and came to attention, and saluted.

The Gunnery Sergeant in charge of the formation saw the old Marine rise and begin his salute. He knew the old man was in his final days. The Gunnery Sergeant barked, "Detail, halt!"

He then executed a right face and brought his right hand to his temple, and held it there.

The old man finished his salute and remained at attention until a second later, when the Marine finished his.

A quick left face and a loud, "Detail, March!" and the Marines were gone.

The old marine said to no one in particular, "I can go home now."

They found him the next morning, a smile on his lips, and cold to the touch.

A Marine had gone home.

WE SAVED THE WORLD

She was a world-class physicist. Weapons weren't even her specialty. She took the job at the weapon's research facility because it paid well. It turned out, she had a natural ability with firing sequence codes. The pinnacle of her career, was the night she broke the Russian nuclear weapons firing sequence codes. Actually, it was worthless information by itself, unless of course, you had a stolen Russian nuclear warhead that was already armed in front of you, which she had. It was going to go to Washington D.C., it was going to Tel Aviv. It didn't matter where it was going to go—all that mattered was that if it went off here, it would be seen as the U.S. government using a nuclear weapon against a peaceful people; and that would start a war that would never end.

They were a half-mile underground in an Islamic terrorist hide out. The way in was littered with dead bodies and empty M-16 magazines. She should be at the mall, shopping. The weapon was armed and counting down. There were dead Marines all along the route they had taken in. There were an even greater number of dead terrorists in various side tunnels.

She was in great shape. That was one of the reasons she was asked to go on this mission. She'd had her hair done last week; it needed to be done again, soon. There might not be a soon, if she couldn't stop the firing sequence, there might not be tomorrow for anyone.

Anywhere.

It wasn't her fault; she just went to work one day and this man was there. He lied about his name. She knew that now. Not that it did her any good in her current circumstances. He had talked about lives. He had talked about obligations, he had talked about country. If he had talked about duty and honor, she would have said no. She didn't know about such things then, but she did now. She had seen good men die for duty and honor, and it was all for nothing, because she couldn't stop the countdown. Someone better than her had found the code and changed it. In five minutes they would all be dead, because she couldn't do what they had died to bring her here for. The

CIA agent, she, and a Marine were all that were left. She turned to the Marine and, said, "I can't stop it. I can't stop it."

The Marine was a man of few words. He simply said, "You really need to stop it now."

"I can't!" She screamed. "They changed the damn code! I could do it in a year. I could do it in a month, but I can't do it in five minutes!"

Then I'll do it in four minutes." The Marine said, as he slapped a device onto the warhead and flipped a switch.

"What the hell did you just do?" The CIA agent asked.

"That's five pounds of C-4. It goes off in four minutes." The Marine said. "If we're lucky it will blast the bomb apart and it won't go off. In any event, this whole cave system will come down on top of it. We should start running now."

A half-mile in four minutes, she could do. She had done it many times. Not after a forced march of twenty miles in two days. She hadn't done it after a death filled fire fight into a terrorist stronghold. She hadn't done it after the greatest single failure of her life.

"I can't," It was all she could say.

The Marine grabbed her by the shoulder and threw her towards the exit. "Move now!" He barked.

Something inside her responded to the tone of command in his voice and she started to move, just not fast enough. The Marine butt-stroked her with the M-16. "Faster!" He yelled. She got a lot faster after that, because that is when the gunfire started.

Every cross tunnel they came to, gunfire erupted out of. She ran faster. The Marine and the CIA agent came behind her, firing as they ran.

On the best day of your life, you can run a half a mile in four minutes. On the worst day, you can do it in three-and-a-half. She ran faster. They shot more. The good part was, she forgot about the nuclear warhead behind her counting down to Armageddon. She forgot about the five pound of C-4. She forgot about anything but running for her life.

The detonation, when it came, dropped the mountain on top of the tunnels. It buried the radioactivity for ten thousand years. It erased any trace of the terrorist stronghold It erased any trace of their operation; but it didn't erase them. They made it out by a hundred yards.

"We did it!" She yelled, slightly delirious. "We did it! We did it!"

She looked at the Marine, "Now what do we do?"

"We walk back to the pick up point." He answered.

"I'm not walking back." She said calmly. The Marine had already started walking. She ran to catch up with him. "I want a helicopter." He didn't respond.

"I want an SUV or armored personnel thingy! I walked in, I am not walking out!"

The Marine stopped and looked at her, "Okay," He said. "Then you die out here. Nobody is coming here to get anybody."

It only took the government agents a couple of weeks to get her to agree to never talk about the operation. That, and a few million dollars they offered to put in her bank. Then of course, the threat of her spending the rest of her life at a safe, secure, undisclosed location, for the good of the country.

Funny thing, she thought, those government agents, the ones in the suits, the ones who had never worn camouflage, they never talked about duty or honor. Those were the reasons she finally agreed to stay silent. She learned those concepts from the Marines who fought and died that day. She learned that from the men she had looked down on when she first met them. She learned that from the men who died to save the world. She learned that from the men she would love for the rest of her life.

Every year after that, a black limousine pulled up to the gates at Arlington National Cemetery, and a woman walked the grounds, and prayed and wept.

Happy Veterans Day.

Happy Veterans Day to all of us who, at one time or another, saved the world.

JUSTICE

His father had always been the best man he had ever known. He had taught him how to hunt and fish. They had skipped stones at the lake. They had swam in the ocean. His father had taught him to respect nature and himself. One more thing, the most important thing of all, his father had taught him to never hate, "Hate," His father had said, "Is a destructive emotion. Over time, it rots away at your soul, and reduces you as a human being. Instead, feel sympathy for those you would otherwise hate. Reflect on the poor quality of life and the harshness of their upbringing. Realize that you have been given all the good things in life that they missed out on. That will keep you centered, and in control of yourself, and make you a better person."

He had carried those lessons with him all of his life. He appreciated the good things in his life. He had a beautiful, caring wife, a great son who would start college soon and one of the best jobs in the world. He was grateful for the teachings of his father, and missed him.

His mother and father had finally taken the honeymoon he had always promised her. A first class trip to the Bahamas. They stayed at the best hotel. They even went paragliding; a bit extravagant at their age, but on the phone, his mother had said she was perfectly happy. That just made it hurt all the more when he got word that the hotel they stayed at, had been brought down in a massive terrorist attack which killed hundreds.

It had been so hard not to hate. But he centered himself and remembered the things his father had said. He felt sympathy for the men who had such poor lives that they felt they had no other option than to take the lives of innocents.

He felt honest regret that their lives had been so poor, and wished it had been otherwise.

So, he did not hate, and he was a better man for it. He did not hate as he nosed over his F-117 A Stealth fighter bomber to begin his bombing run on the safe house where the terrorists who had killed his parents were hiding. He did not feel hate as he released the two five hundred pound bombs from the internal bomb bays. He did not feel hate as he watched the house totally and completely obliterated in two massive explosions in his rear display monitor.

He did not hate; but he did smile.

He keyed his radio and reported in, "Overlord, this is Justice. Mission complete. Target destroyed. Returning to base. Out."

OPERATION HERO FLIGHT

As a general rule, the cooler the operation's name, the more difficult the actual accomplishment of it would be. Some genius back at headquarters, had decided to call this one, "Operation Hero Flight." The Marines in the back of the C-130 had never done anything like this before, and they hoped to never have to do it again.

The aircraft hit the runway hard and fast. It was going to be a short landing. The co-pilot looked out the windshield, "There's a crowd just off the runway." He told the pilot.

"There's always a crowd." The pilot said as he clicked on the intercom to talk to the Marines in the back of the aircraft. "We've got company up here people. You might want to get your "A" game on."

As the C-130 approached the end of the tarmac, very close to the crowd of people, the two men in the cockpit could see many in the crowd had their fists in the air, and they were chanting. As the aircraft stopped just across from the crowd, separated only by a gate manned by two unarmed guards, the pilot began to turn the aircraft. "Ahead on Starboard engine, back on port, drop the cargo bay ramp. They want Marines, let's give them some."

In the back of the aircraft, the Gunnery Sergeant walked out onto the cargo ramp before it even hit the ground. He held onto a strap with his right hand and looked out at the crowd. He smiled. He looked back at his men. They had come a long way for this. As the engines throttled down, he raised his left hand, and just as the cargo ramp hit the ground he called, "Marines!" He brought his left hand straight out towards the crowd and watched as the men hit the deck at a jog.

The two guards suddenly opened the gates and ran for their lives. As the crowd surged forward, with the roar of the engines dying down, the Marines could hear the crowd chanting.

They were chanting, "USA! USA! USA!"

Then the Marines had something happen that hadn't happened during their entire year long mission. They were overrun.

They were hugged. They were back slapped. They were kissed.

They were home.

Still on the cargo ramp, the Gunnery Sergeant said quietly to himself, "Welcome home, men. Welcome home."

STREET PATROL

The young soldier had nicknames for all the kids they saw while on patrol. The rest of the guys in his unit called him "Candy Man". For the candy he always handed out when they were on the street. In the midsts of the high tension of their work, the smiles on the kid's faces were a welcome diversion.

It was just like any other day. The soldiers suited up, gunned up and hit the streets to "go see the locals", as their sergeant liked to say. The group had been lucky. The action they had seen had been light, and half hearted; a shot or two here and there. Once, they had actually found an IED, but it had been a dud. Not bad for a deployment to a war zone.

The foot patrol turned onto the street where the boy named Hazzim usually came running up to them.

But instead, he was across the street, just sitting, arms wrapped around his knees. The young soldier looked him in the eyes. The boy's eyes were full of fear and he just shook his head. Then, that quiet little voice in the young soldiers head started talking to him, "We are going to be ambushed."

"Sergeant!" He called. Then he gave the code word for an imminent ambush. "Look around." He said, to prove what he was saying. The street was quickly becoming deserted.

The Sergeant turned to his radio operator and said "Air." The radio operator keyed his microphone and rattled off a string of code phrases and call signs as well as their position.

"Advance and cover!" The Sergeant said. "Let's go find the party."

The soldiers advanced in short bursts from one "hopefully" covered position to another. It's hard to find good cover when you don't even know where the fire is going to come from. The most likely direction is from the front.

The street was a dead end, and as they approached the building at the very end, it started. RPG and machine gun fire erupted into the street from all levels of the structure. The men had good cover and returned fire. The radio operator kept up a steady stream of position and action reports and yelled "Air inbound!"

The Sergeant directed his men well and calmly. This wasn't his first time at this particular rodeo. The tough guy from New York City fired in controlled bursts and prayed silently between trigger pulls.

Overhead, the two Apache helicopters came into view and released a salvo of missiles. Then followed up with 30 mm cannon fire. It was over in less than a minute.

As the soldiers walked back the way they had come, the "Candy Man" noticed a door slightly ajar, and just about waist height, a single dark eye peered out. The young soldier stopped just at the door and knelt down to retie an already perfectly tied boot lace.

When he walked off, a bag of candy lay on the ground. One small hand reached out from the almost closed door, and the candy disappeared into the house, the door closed once again.

THE SNIPER

The mission brief had left no room for questions. The mission was of high importance; and so far into enemy territory, the team could not be seen. And if they were seen, no one who saw them could be allowed to live.

"Gentlemen," The General had said, "If this goes right, this one mission could end the war. If you are found out, it could start a war on a third front, a war that will escalate to unacceptable levels."

To the sniper, deep in the desert with his team, the mission had already accelerated to an unacceptable level. The shepherd had shown up with his sheep about two hours ago. The team had a good position in a cave on a small rise, well hidden, unless of course, this guy walked right up on them, as he was doing now.

At half a click, the sniper, looking through his scope, realized that the sheep herder was unusually short; at two hundred fifty yards he started praying. The shepherd was a boy about ten or twelve years old. Old enough in this part of the world to be responsible for his families herd. He was old enough to die.

The sniper had no children. He had never had a successful long-term relationship with a woman, and hadn't visited his family in years. He could not recall ever having been in a church, except for one time when he was surveiling a target. The team was his family, his work was his life. The nicest thing you could say about him was that he was a really good killer.

The really good killer started to pray. He didn't think or speak in complicated rhythms. He said what he had to say, and that was all. His prayer was that way. "Please God. Please God. Please God. Make this kid go away."

The boy moved closer, looking at something on the ground. He was chasing a lizard. He was playing.

"Damn it!!"

"Please God. Please God. Please God." The sniper's trigger finger moved inside the trigger guard, not touching the trigger, just sitting there, close to the trigger.

The boy moved closer, following the small lizard deeper into the danger zone, closer to death.

"Please God. Please God. Please God." His finger curled around the trigger. He blinked to clear the sweat from his eyes. "Please God."

The boy started to look up, this close he would see the sniper. No one who saw them could be allowed to live.

"God! If you're there, HELP ME!" His finger touched the trigger very lightly. He drew in a breath and began to exhale slowly. He could see a small mole on the boys left cheek.

Something rustled in a bush to the boy's right. He looked that way, and suddenly a bird took flight. He turned to his herd, and went back to the camp. He walked away from the sniper, away from death, back towards life. He would go to college. He would marry a good woman and have children. He would be a doctor, and do good works. And under no circumstances could anyone ever be allowed to know where money for his college fund came from.... ever!

The mission came up dry. The target wasn't there. Not the first time that had happened. When they got back to base, the sniper said to his C.O., "I'm done."

The sniper would not go to college, he would not marry, and he would never have children.

He did spend the rest of his life working with disadvantaged youth, helping them stay off the streets.

A small sign on his office wall read simply, "*What you do for the least of these, you do for me.*"

A SNIPER'S TERRITORY

The man who was once known only as "the sniper", had gotten on a random cross country bus, and got off at a random spot. He didn't know the name of the city, he was only dimly aware of the state he was in. The only thing he was sure of, was that in the desert, on the day that had changed his life, there had not been a bush when he had put his sights on a ten year old boy. Before it was over, the bush was there, and he didn't know how.

He had been born to a woman who was more whore than lady. She had an endless string of boyfriends, who never seemed to find any problem with slapping around a young boy. She never even complained. His survival had been his own concern since he was a very young age. When he showed up at the Army recruiters, the knife in his pocket still had the blood of her latest boyfriend on it. He had taken care of himself. He was well prepared phsycologicaly; well prepared for the hard jobs the Army would ask him to do.

He turned left and began walking down the street. His army issue bag over his shoulder, the rest, he left to chance. As he was passing a fifteen passenger van, a ladder fell off. He easily caught it and let it down to the ground. A priest came around the corner of the van falling over himself apologizing.

"No problem, Father." The man who used to be a sniper said with a smile, "Where can I put this for you?"

"Just inside the cafeteria would be fine." The Priest answered. "Now that school is out, I thought the place could use a coat of paint."

Another ladder, several five gallon cans of white paint, various brushes, paint pans, and bundles of rags later, the priest introduced himself as Father Michael. The ex-sniper looked at the brush in his hand, and introduced himself, "Painter, John Painter."

With a strong handshake, he had a job. He could not only produce his own driver's license and social security card, but by the next morning, the state data bases would be changed to agree with this new persona.

Over the years, he had used many names. He had access to passports from several countries, and changing identities was as comfortable for him as changing shirts. His childhood had been an unending nightmare of abuse and constantly moving from place to place, because his mother was wanted in more states than she wasn't.

He had joined the army strait out of high school and found that the more dangerous a mission was, the better he liked it. He was quickly picked for Special Operations. He had two Military records. One record was for public consumption, the other was locked in a safe, inside a secure facility. He not only knew where the bodies were buried, but he had put many of them there personally. Credit cards and a new military history would be procured by the end of the week.

John had already decided to settle here. He had many resources. Although he preferred jeans and a well-worn button-down shirt, he fit in with the area. In a week, he would have a ware house and arrange for his things to be delivered. He had accumulated an interesting collection of mementoes over the years. Most of them, if handled correctly, were capable of producing great damage. That was the primary reason he had set them aside before he retired. Along with several millions of dollars he had acquired, by mostly illegal means during his missions.

He rented an inexpensive apartment in a seedy part of town, where people didn't ask too many questions. Weapons, drugs, and cash all moved easily. He had no use for the drugs, but he was comfortable there.

John Painter found a friendly bar within walking distance of his apartment. Eddie's Bar. Here, he began to create his legend. A legend is the story intelligence operatives build and share freely. A legend isn't the truth, but it is useful.

John picked out his bar stool so as not to have his back to the door. He played pool well, being careful not to win too often, but always willing to put a few dollars on a game, just to be friendly. He liked his beer cold and his women hot, although he never seemed to have too many of either. He talked hunting, fishing and sports. He was always willing to help you fix your car. And if you were down on your luck, he was good for a beer or two. He took the good natured ribbing about working for a Priest, with a smile. In short, he had a bar; where if the cops asked questions, he was just a good guy.

John quickly became the church's handyman and all around go-to guy for special projects. The church had a school with an after-class youth center. Father Michael had high hopes that participation in after-class activities, would bring higher achievements for his kids. There was just one problem. The drug dealers were moving onto the playground. What they didn't know was they were moving onto a sniper's territory.

One of the students, a pretty young black girl named Christina, poked her head into John's office at lunchtime and said quietly, "They're here."

John looked up from the papers on his desk and asked, "How many?

"Three" She answered.

"Easy day," The Sniper answered, "Go to your next class. This conversation never happened."

The girl left, the killer reached into his desk drawer and pulled out his Colt 1911, 45 caliber semi-automatic pistol. He didn't like to be too far from a fire arm. He racked up a round and took the weapon off safe.

The weapon tucked under his shirt, he went out to his car. The action he had decided would take place off of the school grounds. He watched the drug dealers get back in their black SUV. He cut them off after half a block. The SUV screeched to a halt. John was at the door almost before it stopped. He opened the door and pulled the screaming driver out before he could react. Two quick throat strikes and that one was down. John took the pistol from the dealer's waist band and shot the other two who had already gone for their guns. He caught a round in the shoulder for his trouble. Luckily, it hit no bone and missed the muscle. John patched it up himself at his apartment, and still managed to make it to Eddie's at his usual hour.

The SUV was found in the river a few days later. The cops called it gang-on-gang violence and went back to eating doughnuts. Life is cheap in the city.

The next day someone in a silver Cadillac sprayed the school yard with gun fire. John decided it was time for the dealers to relocate; to hell.

The drug dealers had a house downtown that they had transformed into a fortress. Steel security doors, bars on the windows, and lots and lots of automatic weapons. It was impervious to any attack the cops might mount. Too bad for them John wasn't a cop.

The killer had a first rate security system on the warehouse. He entered the code on his phone and drove in after the garage door opened. He parked his car and walked over to the trailer he used as weapons storage. It was time to select his tools

The killer was in full charge of his brain now. Left behind was the silent sniper. Completely absent, the crying, defenseless, young boy. This new persona was a showman, a master of death and destruction, soulless, pitiless; he would have his revenge. He would keep his chosen flock safe, and no power on earth could stop him.

This newly created persona reached for the M-79, a single shot 40 mm Grenade launcher and smiled. He grabbed a belt of High explosive rounds, and a belt of incendiary rounds. Nothing could resist the phosphorescent warheads burning at 4,000 degrees. He put these two items on the passenger seat of his car.

He returned to the trailer and looked around him. He could hold off a small army with what he had in there, but he meant to destroy one. His hand reached for one of the claymores lovingly displayed on the wall. He checked the safety, and pulled on the trip wire, it was still good. This went into the trunk of the car. On one side in raised letters were the words, **Front; This side towards enemy**. He made sure that side faced towards the back of the trunk.

Again, he returned to the trailer. He spoke aloud; almost reverently, as he opened a cabinet at the back of the trailer, "No party would be complete without the AK-47, one of the sweetest killing machines ever made." He locked the slide to the rear and ensured the chamber was clear. He grabbed a green shoulder bag with ten fully loaded magazines. He selected a magazine and inserted it into the receiver. He gave it one good slap to insure it locked into place. Gun and bag went on the back seat. His .45, he placed in its shoulder holster, and the killer was ready to play.

The dealer's house was at the end of a dead-end street. The houses on either side had been burned to the ground. Behind the house, there was nothing for miles. They really had made the job easy for him. Isolated target, no chance of collateral damage, and no way out.

It was just past midnight, the party was in full swing. He had already placed the claymore with its tripwire, across the back door. The killer didn't bother with subtlety. The stolen car he was driving just pulled up and stopped a couple of empty houses down the street, and got out. He opened the action on the M-79 and inserted a High Explosive Round. He closed the weapon, took aim at the High security door and fired. He threw the M-79 back in the car and picked up the AK. He liberally sprayed the front of the house with the weapon on full auto. He put in a fresh magazine and waited. It was now a two story house with no front door. The killer could see from closet to staircase. If the first explosion left any doubt, the machine gun fire made it very clear they were under heavy assault.

The AK he switched to single fire as some people tried to get out the front door. He picked off the first few who tried. Like rats in a maze, they learned quickly and headed for the back door. A few men tried to return fire from various windows. They died quickly. The tripwire to the claymore was activated by the first of many people who crowded the downstairs hall. They turned into a pink cloud when the Claymore blew. Smiling widely, the killer switched back to the M-79. He loaded a phosphorescent round and sent it into the middle of the now-wrecked structure. The morning sun would find the building burnt to the ground and no survivors.

John wasn't even breathing hard as he drove off in his stolen car. He had to pick up his own car and clean his weapons. The killer, his bloodlust satisfied, went back to the dark regions of John's brain, waiting patiently should he be needed again. John's kids were safe, and they always would be.

The police investigation was thorough, but useless. All of the ammo had been stolen from lot numbers that had been stolen themselves. Clues would lead all over the world, but none could be traced to John. He got his explosives from dead men.

The kids at the youth center were overjoyed that they finally beat John at basketball. John of course, blamed it on a sore shoulder he got from bricklaying. "Getting old, Johnny!" They teased him, "Come on back when you want some more." It was a good confidence builder for the boys. John of course, wanted only good things for "his kids".

HONORS GUARD

It was Honors choice,
Which laid him low.
Honors choice,
Which struck the blow.

It was honors choice,
To stand and fight,
When others ran,
Into the night.

It was honors choice
To make the call.
Stand your ground.
We fight for all.

Honors choice,
A sturdy thing,
Which made men stand
Beside their king.

Honors choice
Where freemen dwell
Cast the demons
Back to Hell.

They brought him home
To widowed queen
And a son
Not yet a teen.

Honors choice,
To watch the yard.
They were called,
The Honor Guard.

BOOT PRINTS IN THE SAND

The young Marine hit the ground hard, bullets kicking up sand all around him. His pack was heavy and his M-16 brand new. It was his first time in-country, and the first time men were actively trying to kill him. He landed in a low spot, next to an older grizzled veteran. Unshaven, his BDU's (Battle Dress Uniform), torn and dirty, eyes dark and hard, he smiled. "Congratulations kid." He said, "You're a hero!"

Something didn't make sense, the young Marine thought. He hadn't fired a shot. He basically fell, almost on top of the old Marine, just a few yards from the vehicle, and this guy called him a hero. Convinced he had heard wrong, he yelled, "What?"

"Look behind you." The older man said, "Those boot prints in the sand, they're yours. You're a hero!"

The young Marine put a few rounds down range and thought to himself, sarcastically, "Great, I made it."

Years went by, and the young Marine, now a grizzled veteran himself, heard "Taps" in the distance and thought to himself, "God, I hate Taps. I have always hated Taps. It means another hero has gone on." Then he heard a familiar voice, "Congratulations kid you're a hero!"

He was instantly taken back to that day, so many years ago. The first time he heard those words, the day he had helped bury the man who said them, and he saw that same face again, and that same crooked smile. His response was the same, "What?"

"Look behind you. Those boot prints in the clouds, they're yours."

CHAPTER 8. Mission Ops

Mission Ops embody the warrior's heart and soul. A duty to protect and honor those that are in need, and honor those that have fallen before them. Here are a few final stories, although they are fiction, are accurate in detail, as far as weapons and tactics go. In the dark and dirty world of Special Operations, when teams are inserted behind enemy lines, they often have only each other to depend on. They are in very good company. This is one last salute to the warriors of the darkness with whom it was a great privilege to serve......

A DRINK WITH HIS FRIENDS

Two army Chinook Helicopters were called in for an emergency evacuation. The patrol was surrounded and pinned down on a ridge line south of Kahndahar; taking heavy casualties, the Lt. in charge decided it was time to get out of dodge.

Two F-18's came screaming in and bombed the nearest enemy forces to knock them back long enough for the Chinooks to land and load up the troops. One Chinook took an RPG in the engine two hundred yards from the LZ and crashed in a fiery explosion. There were no survivors. The remaining Chinook came on, the pilot determined to get the soldiers out.

Sergeant (Sgt.) Miller, the men called him Sgt. Rock—because he loved the action, and wasn't afraid of anything—got a really bad feeling in his gut as he saw the first Helo go down. Now, there was not enough transport for everyone. He ordered, "Load the wounded first!" Bob, Pete and Chris were his perimeter force; they had volunteered to give cover-fire during the evac. They were good soldiers. They held the enemy at bay and saw the crash; they knew what was coming.

Sgt. Rock had to physically throw two soldiers onto the aircraft because they refused his direct order to board. He wanted everyone off the ridgeline that could get off. The door gunner yelled. "That's all we can take! We'll come back for you!" Anymore weight, and the Helo would not be able to take off. He locked eyes with the LT who had been one of the first to board. The LT looked at the young Sergeant with horror on his face, and then he looked down in shame. The Sergeant reached in and grabbed the seriously wounded private Dave Engles and yelled in his ear, "Arlington on Veterans Day! Arlington on Veterans Day!" Unsure if the private had even heard him, he backed away to let the Helo lift off. As the Helo cleared the ridge, he heard Chris yell, "Here they come!" Moving at a trot towards his men the Sgt yelled, "Give 'em Hell!"

Dave Engles was too badly wounded to remain in the Army, so after his rehabilitation, he opened an ice-cream shop. He had developed a limp from combat wounds. It was somewhat painful, but he told his girlfriend that the worst thing was, he could never stand at full attention that he felt he should, in order to salute the flag. He had gone to Walter Reed, where several specialists saw him. They did what they could for him and sent him home. Two of the doctors spoke after he left. The younger of the two said, "There is nothing wrong with that man's leg."

The older doctor looked at the younger and said, "The problem is not in his leg, it's in his head. I read his whole file, which by the way, you should start doing. Psychological trauma due to combat."

Dave met Marlene at the Veterans Assistance Center. She helped with the paper work for his loan to open the ice-cream shop. They fell in love over hot fudge Sunday's. As the business grew with inventive menus, and colorful paint jobs inside the shop and out, Marlena suggested he open another shop across town. She was a smart businesswoman, and knew a good idea when she saw it. He said he wasn't ready for that just yet.

The following November, Dave told Marlena he would be going to D.C. to "see some army buddies." He wanted to see his friends before he made a decision. Even though she asked, he would not tell her what the decision was about.

As the day approached, she insisted on helping him with his plans. They double-checked the bus route, since the war, Dave really didn't care to fly. He got on the computer and printed out a map of where he was going to meet his friends. Marlena asked where they were going to meet, and he folded the paper and smiled as he said, "Military secret."

"Yeah right, "She replied, "Secret Military night club." And laughed.

The last thing he packed on the morning he was to begin his long bus ride, was a whiskey flask. "When we knew it was gonna be bad, we would drink a shot. You know, just in case. I carried the flask. First to drink was Sgt. Rock, then Bob, followed by Pete then Chris; I would drink last and put it back in my vest pocket. The joke was I would live forever because the bullets would just bounce off the flask."

That was about as much as he ever said about the war. Working for the VA, Marlena knew many soldiers preferred not to speak about their time in combat.

As Dave got on the bus, he turned and kissed her; he looked deep in her eyes and said, "I love you. I'll be back."

Something wasn't right, she decided. He was kind of sad, not like a man going to see his buddies. She decided to have a little talk with the SgtMjr who had taken a liking to her. He was in charge of records. Maybe he could tell her something about the guys.

Dave took a cab from the bus terminal. He got out at the gate that led into Arlington. He passed the guard at the gates, where a young soldier, looking good in his dress blues, a hand on the ceremonial sword at his side, stood guard. Some women and children stood a little ways behind him.

Above the gate was a sign inscribed with the words:

ON FAMES ETERNAL CAMPING-GROUND
THEIR SILENT TENTS ARE SPREAD
AND GLORY GUARDS WITH SOLEMN ROUND
THE BIVOUAC OF THE DEAD

Dave took the "Secret" map from his pocket. He was to go one-half mile in and take the trail to the left. He knew he would see a large oak tree, then twenty-five paces to the southeast he would meet his friends. With his limp, it was painful, but he took a pill and carried on. The pain was worth it to visit his friends.

As he passed two soldiers in cammies, the older one grabbed the younger soldiers shoulder and said, "Congratulations kid. You're a hero."

Dave continued on and took the left fork. He passed a father and son. The father was crouched down to his five-year-old son's eye level and said, "Never hate, son. Never hate."

He passed a group of Patriot Guard riders in their motorcycle chaps and vests. The oldest of the men seemed to be in mid-story speaking to the rest. He said, "Yeah, he was a good soldier, he always stood his ground."

A bit further on, a family group seemed to be having an argument. The grandfather in the wheelchair was saying, "I am going to the parade. What's it going to do, kill me?" That seemed to end the discussion.

He was passed by three heavily muscled men. Dave saw that the youngest man's hands were healing from heavy cuts. He didn't know how the younger man's hands got so badly cut, but he knew that must have hurt like all get out.

He came to the Oak tree and turned left onto the grass. "Twenty five yards southeast." He said to himself. His hands started to shake, he put the map away. His leg ached. He limped on.

Twenty-three, twenty-four, twenty-five paces. He stopped. Exactly as he had been told, there were his friends. He spoke quietly, "Hey Rock, Bob, Pete, Chris." His voice choked up a little, so he cleared his throat to speak. "Sorry it took so long to get here. Rehab was rough. Thanks for watching my back over there."

He went on to tell his friends about his ice-cream shop, and about Madeline. He had fallen in love and wanted to ask her to marry him.

Thirty yards away, a young black boy tugged on his father's sleeve and pointed."Daddy, why is that man talking to himself?"

The boy's father, a tall, muscular man, wiped a tear from his eye and said, "He's talking to his friend's son, and I can just about see them." They had come to visit the boy's grandfather's grave at Arlington National Cemetery.

Dave fell to his knees and cried out, "I'm sorry! God I'm so sorry! I didn't mean to leave you guys there like that." Then he cried as he said, "Forgive me. Forgive me." Dave felt their strong familiar arms around him. Comforting him.

He heard Sgt Rock say, "Nothing to forgive, Indiana." They always kidded him about being from the state of Indiana. "Not a problem," Bob said, "We knew what we were getting into."

"Glad you got out in one piece." Pete said, "Somebody had to tell the tale."

"Don't worry about it knucklehead." Chris said, "You still carry that flask?"

Dave stood to his feet and wiped the tears from his eyes. He took the flask out of his pocket and poured a shot on each grave, in the proper order. First, Sgt Rock, then Bob, and then Pete, then Chris, and finally, he raised the flask and said, "Thanks, guys. I love all of you."

Then he took his shot, replaced the flask in his coat pocket, came to full attention and saluted his friends. He did a quick about face, and marched away. His limp gone, his pain had been his guilt. And guilt goes away when your friends tell you that you are forgiven.

In the tree line ahead of him, he saw Marlena. They hugged and kissed and he said, "How did you know?"

"Wounded Warrior, Washington, and Veterans day?" Marlena said. "How could I not know?"

Then she realized, "Dave, you're not limping!"

"I know. I talked to the guys, and now my leg doesn't hurt. How about we fly home? We have a new shop to open and a wedding to plan. I don't have a ring right now but..."

She interrupted him, "Yes! Yes! Yes!"

He was almost knocked off his feet by the hug; but then, he didn't have a bad leg anymore.

STRIKE TEAM DAGGER

Assad, also known as "the dark one", had always been a good Muslim. Bearded, as a good Muslim should be. He was an even better fighter. Assad enjoyed seeing the light go out of the eyes of the infidels he killed. He had been chosen by the Imam to be one of the guards of this most valuable of sites. The infidel archeologist that had found this holy site, had died slowly at his hands. Because he was so devoted in his following of the prophet, Assad was allowed certain latitudes in his treatment of any infidels who came his way.

The trackless desert has no landmarks, only a select few knew how to safely cross the wastelands. Assad was on guard, and though there were no infidels nearby, he was happy because he was serving Allah.

From his post at the entrance to the cavern, his view was restricted due to geological formations. The cavern was situated inside the base of a mountain. A narrow but deep depression led to the first of many caverns and numerous tunnels. The entire area was dotted with rocky outcrops, which erupted from the ground and rose for forty or fifty feet, and could stretch for miles. You didn't need the cave to hide an army here, but it was helpful.

Unknown to those on the ground, a Global Hawk had been making close observations of the area for several days. It had followed a supply convoy from across the border in Pakistan. A lucky break for the infidels, not so good for Assad and his brethren in arms. Assad, happily ignorant that their position had been compromised, was actually smiling as the blade of his enemy severed his carotid artery and then his spine. The leader of Strike Team Dagger laid the now dead Islamist gently on the ground. Dagger had briefed two days ago. Their mission: Parachute in and rescue a Nobel Peace Prize winner, Professor Wellingsworth, who had foolishly gone into the desert without a security detail. He had been taken hostage on his first day. Now dagger had to go get him out. At thirty thousand feet, the Global Hawk silently recorded and transmitted the images around the world.

Movement in the darkness, a silenced shot and the high guard was down. Jeremiah spoke softly, "Scabbard, Dagger we're in." The five-man team moved swiftly to occupy the strategic points at the mouth of the cave. They stopped and listened for any sound which would indicate they had been heard.

Mac, the heavily muscled red-haired explosives expert, started laying his charges. Preprogrammed to explode in a specific delay pattern, they were meant to give the team a chance to clear the entrance. Some men are demolitions experts and some are artists. Mac was an artist. The team had once escaped from a grim-looking shootout using just what Mac found in the kitchen. The explosion killed four of the main attack force, and the team got out undamaged.

The main cavern, some two hundred by one hundred feet, was dimly lit. Tunnels to the left and right, led deeper into the earth. Foot steps and talking from the tunnel on the right. Two silenced pistol shots stopped the footsteps. The team moved to the right and posted Private Brown, nicknamed Slim for his slight build, as rear Security. Brown had worn out his welcome with the crime families in New York City. He had found a wonderful home in the Marine Corps. After all, he was a good killer. Even though he was an excellent fighter, the junior member of the team, he had yet to earn a spot on point. His day would come.

Some two hundred yards inside the mountain complex, Professor Emanuel Kent Wellingsworth lay dying. A Noble Peace Prize winner, he had gone into the desert in an effort to find peace for the region. His guide led him to the mud brick house where he had been taken prisoner. He had since been questioned, beaten, traded between groups, and finally arrived here in the night. With no answer to their demands for ransom, it was decided to ignore him and allow him to simply expire without interference.

Gunfire was a common enough occurrence inside the mountain. On this particular occasion, it sounded heavier and more sustained. The Professor crawled deeper into his cell for protection from the rounds that seemed to be coming closer to his cell. He had long ago given up hope of ever leaving alive, now he was simply trying to stay alive for another hour; or perhaps minutes. These men were not the educated and thoughtful leaders he had hoped to meet months ago, when he began his self

-appointed peace mission. He'd had time to get to know these men and had concluded they were little better than thugs. This moment he came to believe was his last of life. As an avowed Atheist, he had no God to make peace with and his mind was rapidly deteriorating.

In the darkness, an explosion rang out. This was the generator being destroyed. The lights went out and the team was happy. The team's long familiarity with night vision gear gave them a wonderful advantage in the dark. The team ran into a stubborn pocket of resistance at one junction where two tunnels met. Cherokee, who actually was a full-blooded Cherokee Indian, worked his way to the junction under cover fire from the rest of the team and chucked a fragmentation grenade around the corner. The blast so close to his cell, it damaged the Professor's eardrums; and his sense of balance. He saw vague shadows at the bars of his cell, and hoped they would pass him by. Wild-eyed and fully bearded, being behind bars and unarmed probably saved his life.

The team formed up into a closely packed line and came around the corner; just past the pile of dead Arabs, they found their prize. Cherokee took station just past the cell to provide cover fire with his M-4, while Mac set a carefully calibrated explosive to blow the lock. He pointed at the terrified man in the cell and covered his ears. The professor did likewise and the lock was blown.

Mac and his longtime friend, Sgt Davidson rushed into the cell and got "the package", Professor Wellingsworth, ready for transport. When the commandos started to put the bullet proof vest on the man, he struggled weakly. The Professor lived in constant fear that he would be used as an unwilling suicide bomber, and now his worst fears seemed to be coming true. He was struck in the face once, and ceased resisting. A hard shell placed on his head, to blind him, he assumed, finished the preparation.

The strike team withdrew at a fast pace but in good order. Jeremiah, whispered into his microphone, "Dagger five. Dagger is outbound clean." This let Brown know that the team was coming back out with no hostiles in pursuit.

The Professor was jerked roughly along corridors and around corners. At one point, he was thrown to the ground, and a heavy weight landed on his body knocking the wind out of him. Completely confused and terrified, he knew he would soon die. The weight came off of him and strong hands under his shoulders, lifted him to his feet. Again, he was moved along the corridor. A burst of gunfire sounded, and he was slammed against the cave wall. In the flash of light from the gunfire, he saw several men, shadows in the darkness. He heard gunfire ahead of him, then thrown to the ground, a body slammed on top of him, and heard a thunderous explosion. The Professor was in hell.

Private Brown was a little busy at that moment, lighting up the left hand tunnel with fully automatic fire from his M-249 Squad Automatic Weapon (SAW). He responded to Dagger one by yelling, "We have company!"

A forty mm grenade from Jeremiah's M-203 grenade launcher went by Brown's head and took care of the company long enough for the team, with their unwilling guest, to clear the cave entrance, and start out of the cave towards their extraction point. Mac stopped and turned halfway out. He withdrew a small box from a pouch on his belt. He smiled as he turned the dial to arm and pushed the detonate button. Then he turned and ran like hell. The timing on his explosives did not allow for falling down.

Davidson dropped behind some rock formations for cover and took aim at the tunnel entrance. The explosions started. First one, then another closer. Mac came running out at full steam and actually dived over Davidson's position. A running man with a beard, in a turban and robed with an AK-47, came into view— Davidson put a round in his chest. The rest of the explosives blew and the cave mouth collapsed.

Jeremiah got on his radio and sent his transmission, "Package secure. Ready for pickup."

Fifteen minutes later, the MH-47G Special Operations helicopter arrived. As the last man on, Jeremiah yelled, "Go. Go. Go!"

Jeremiah carefully took the Kevlar combat helmet off of the Professor. The helmet was tinted for daytime operations, and in the gloom of the cave, and the darkness of the night, effectively rendered the wearer blind. The professor, dehydrated and starving, could not effectively process what was happening. He still believed himself a prisoner and on his way to a horrible and bloody death.

He refused the canteen, believing the water to be drugged, and looked around at each man confused and fearful. In the noise and vibration of the helicopter, Jeremiah grabbed the man's torn and bloody shirt, and leaned in close to the Professors ear and yelled, slowly and clearly. Receiving no response he tore the U.S. flag off his shoulder and pressed it into the terrified man's hand. One small white light from the team leaders' kit shone on the flag, as he yelled over the engine noise of the helicopter once again, "We are United States Marines! We are taking you home!" He watched the man's face for any reaction. The professor wrapped his hands around the flag and weakly mouthed the word, "Home." Then he began to cry.

Jeremiah slapped him on the shoulder and moved away for the medic to start doing his job. It doesn't do a lot of good to free a prisoner and then let him die.

A few days later, in the relative safety of their Forward Operations Base, which was not marked on any generally available military map, the team watched CNN. The Professor, from his hospital, gave a news conference. He stated he deplored the violence which led to his freedom, as he deplored the violence that led to his capture. Any taking of life he said, he regarded as immoral.

Brown spoke up, "Hey Skipper, ain't that your flag in his shirt pocket?" The whole team laughed uproariously.

A red light in one corner of the room began blinking. Jeremiah stood and said in a voice that was an order. "Gear up and gun up. We fly in five."

ANOTHER NIGHT, ANOTHER JOB.

THEY WOULD BRIEF IN THE AIR.

THEY ALREADY KNEW IT WOULD BE SHORT AND BLOODY.

THAT WAS JUST THE TYPE OF WORK THEY DID.

ONE SHORT EPILOGUE

Thank you for reading *Five Generations, Stories From My Father*. It was created and passed down through five generations of American Warriors. Some spent more time at duty than others, but all with one common denominator:

PROTECT, HONOR AND SERVE.
HELP THOSE IN NEED.
AND GOD'S SPEED TO THOSE IN BATTLE.

Ronald J. Aubrey,
QM1 USN Retired
Special Operations Qualified

Books by Ronald J. Aubrey

Five Generations, Stories From My Father (2015)

Books by Ronald J. Aubrey and Deborah Aubrey-Peyron with artist Shane Christian Aubrey, BA

An Old Man's Christmas
(due December 2015)

Books by Deborah Aubrey-Peyron

Miraculous Interventions (2011)
Miraculous Interventions II, Modern Day Priests, Prophets, Pastors & Everyday Visionaries (2012)
Miraculous Interventions III, 2012
The Miraculous Year (2014)
Miraculous Interventions IV The Gathering
(due Christmas 2015)
Christmas Chaos! (2012)
Christmas Chaos! Coloring Book (2015)
My Faith Journey- Dennis Murphy (2015)
Dave, Let's Take a Walk –
The Pastor David Becker Story
(due out fall of 2015)

www.ingramcontent.com/pod-product-compliance
Lightning Source LLC
Chambersburg PA
CBHW022004090426
42741CB00007B/891